SpringerBriefs in Computer Science

Series Editors
Stan Zdonik
Peng Ning
Shashi Shekhar
Jonathan Katz
Xindong Wu
Lakhmi C. Jain
David Padua
Xuemin Shen
Borko Furht
VS Subrahmanian

For further volumes:
http://www.springer.com/series/10028

Michał Wódczak

Autonomic Cooperative
Networking

 Springer

Michał Wódczak
Ericsson (Telcordia Technologies)
Poznań, Poland
michal.wodczak@ericsson.com

ISSN 2191-5768   e-ISSN 2191-5776
ISBN 978-1-4614-3099-5   e-ISBN 978-1-4614-3100-8
DOI 10.1007/978-1-4614-3100-8
Springer New York Dordrecht Heidelberg London

Library of Congress Control Number: 2012932962

Printed on acid-free paper

Springer is part of Springer Science+Business Media (www.springer.com)

With special thanks to my parents

Preface

Cooperative transmission aims to improve the reliability of wireless mobile communications through the use of diversity provided by additional relays assisting in the transmission between the source and destination nodes. This is possible as the rationale behind spatio-temporal processing can be rather easily and efficiently mapped onto networked systems. Autonomic cooperative networking studies the further evolution of this phenomenon by first involving the network layer routines and then, additionally, incorporating the notion of autonomic system design. This book provides both the description of evolution and incremental analysis of these concepts, starting from spatio-temporal processing and its translation into cooperative transmission in ad hoc environments. Following, routing related enhancements are proposed and the whole solution is positioned under the umbrella of autonomic cooperative system architecture. Then, the investigated concept is extended to autonomic cooperative deployments in relay enhanced cellular systems and, eventually, a new dimension is added by putting it in the context of emergency communications.

December 2011   *Michał Wódczak*

Acknowledgements

This book contains the most comprehensive and consolidated description of the research work performed by the author over past 10 years. During this time he authored or co-authored over 40 scientific publications such as journal papers, magazine papers, conference papers, and book chapters, as well as contributed to over 40 technical reports within European Union FP6 and FP7 projects. In particular, the part related to spatio-temporal processing, conventional and cooperative relaying, routing information enhanced cooperative transmission and relay enhanced cellular systems was done at Poznań University of Technology, where the author was employed as Research Expert and where he was working on his doctoral dissertation. At that time he was full-time involved in EU FP6 IST-2003-507581 WINNER and EU FP6 IST-4-027756 WINNER II Integrated Projects on the topic of cooperative relaying, as well as he served as Editor-in-Chief of NEWCOM Newsletter in EU FP6 IST-2004-507325 Network of Excellence NEWCOM. Following, the part related to autonomic cooperative system design, cooperative autonomic network deployments and autonomic emergency communications was done as postdoctoral research at Telcordia Technologies, where the author has been employed as Senior Research Scientist and Programme Manager. During this time he was full-time involved in the area of autonomic networking in EU FP7 INFSO-ICT-215549 Integrated Project EFIPSANS and most recently he has been working in the field of emergency communications in EU FP7 SEC-242411 Integrated Project E-SPONDER, where he additionally serves as Innovation Manager. Moreover, the presented work is related to the author's involvement in standardisation activities, as he has been serving as Vice Chairman of the ETSI Industry Specification Group on Autonomic network engineering for self-managing Future Internet (ETSI ISG AFI), as well as Rapporteur of ETSI ISG AFI on Autonomic Ad hoc, Mesh and Sensor Networks. Last but not least, the finalisation of this book was carried out at the time the author has been already with Ericsson.

Contents

Acronyms

ACRR Autonomic Cooperative Re-Routing
ACT Autonomic Cooperative Transmission
ACN Autonomic Cooperative Node
AP Access Point
ARP Address Resolution Protocol
AWGN Additive White Gaussian Noise
BER Bit Error Rate
BS Base Station
CAdDF Complex Adaptive Decode and Forward
CDF Cumulative Distribution Function
CDMA Code Division Multiple Access
CFR Chief First Responder
CSI Channel State Information
DAD Duplicate Address Detection
DE Decision Element
DHCP Dynamic Host Configuration Protocol
DN Destination Node
DSTBC Distributed Space-Time Block Coding
ECMP Equal Cost Multipath Protocol
EGC Equal Gain Combining
EDSTBE Equivalent Distributed Space-Time Block Encoder
EOC Emergency Operations Centre
FR First Responder
FRN Fixed Relay Node
FRR Fast Re-Route
GANA Generic Autonomic Network Architecture
GVAA Generalised Virtual Antenna Array
IP Internet Protocol
L3DF Layer 3 Decode-and-Forward
LOS Line of Sight
LSTC Layered Space-Time Coding

MAC Medium Access Control
MANET Mobile Ad-hoc Network
ME Managed Entity
MEA Multi-Element Array
MEOC Mobile Emergency Operations Centre
MIMO Multiple Input Multiple Output
MISO Multiple Input Single Output
MLSE Maximum Likelihood Sequence Estimator
MMSE Minimum Mean Square Error
MPR Multi-Point Relay
MRC Maximal Ratio Combining
MRN Mobile Relay Node
MSB Most Significant Bit
ND Neighbour Discovery
OFDMA Orthogonal Frequency Division Multiple Access
OLSR Optimised Link State Routing
OSI Open Systems Interconnection
PSK Phase Shift Keying
QoS Quality of Service
QPSK Quadrature Phase Shift Keying
RAP Radio Access Point
REACT Routing information Enhanced Algorithm for Cooperative Transmission
REC Relay Enhanced Cell
RF Radio Frequency
RN Relay Node
SAdDF Simple Adaptive Decode and Forward
SIMO Single Input Multiple Output
SISO Single Input Single Output
SN Source Node
SNR Signal-to-Noise Ratio
STBC Space-Time Block Coding
STC Space-Time Coding
STTC Space-Time Trellis Coding
SVD Singular Value Decomposition
TC Topology Control
TDD Time Division Duplex
TTL Time To Live
UDP User Datagram Protocol
UT User Terminal
VAA Virtual Antenna Array

Chapter 1
Introduction

Nowadays, as research is being advanced very rapidly, more and more complex solutions are continually devised. There is, in fact, an urgent drive for the development of a flexible networked system, comprising a number of different transmission technologies and able to self-manage, such that seamless, on-demand service provision could be offered to the end users. This book makes an attempt to present a consolidated analysis of the evolution of certain aspects of such a system. In particular, it provides an incremental description of the development of Autonomic Cooperative Networking, starting from spatio-temporal processing and its translation into cooperative transmission in ad hoc environments. Following, routing related enhancements are proposed and the whole solution is positioned under the umbrella of autonomic system design. Then, the investigated concept is extended to autonomic cooperative deployments in relay enhanced cellular systems and, eventually, it is put in the context of emergency communications.

Looking at the presented concept bottom-up, the rationale behind cooperative transmission is to improve the reliability of wireless mobile communications. It is done through the exploitation of the diversity provided by additional relays assisting in the process of transmission between the source and destination nodes. Such diversity is not available in the case of conventional relaying. However, it is feasible to exploit the diversity for the cooperative case, as the notion of spatio-temporal processing may be rather easily and efficiently mapped onto a networked system. In other words, network nodes may act as the elements of a Virtual Antenna Array and perform the operation of a Distributed Space-Time Block Encoder, as long as tight synchronisation is guaranteed. Cooperative transmission is not only applicable to mobile ad hoc networks, mesh networks, and sensor networks but also involves relay enhanced cellular systems. Regardless the environment, however, in most of the cases, there is a need to answer the question of the selection of the Relay Nodes to be included in Virtual Antenna Arrays. The proposed approach is to assume the employment of specific routing mechanisms for the purposes of gaining access to and capitalising on topology information readily available at the network layer. In particular, the Optimised Link State Routing protocol is used which belongs to the proactive class and is well tailored to Mobile Ad hoc Networks. The obvious advantage of the Optimised Link State Routing protocol is the availability of its inherent optimised broadcasting mechanism in the form of the Multi-Point Relay station selection heuristic. A relevant modification of this mechanism allows for seamless integration of the concept of Virtual Antenna Array aided cooperative transmission into this protocol. In other words, thanks to careful extensions to the Optimised Link State Routing protocol, ensuring its backward compatibility, one is able to exploit the readily available routing mechanisms and information for the purposes of organising the aforementioned Virtual Antenna Array aided cooperative relaying. This approach may be perceived, at least to some extent, as more node-centric and, obviously, it is necessary to look at it from a wider network perspective, where numerous cooperative and non-cooperative transmissions can be going on. For this reason, it is necessary to put the proposed Routing information Enhanced Algorithm for Cooperative Transmission under the umbrella of the Generic Autonomic Network Architecture. This is done through the introduction of the Autonomic Cooperative Node and the incorporation of certain managing entities, such as the Cooperative Transmission Decision Element and the Cooperative Re-Routing Decision Element. Following, the analysis is extended to Relay Enhanced Cell where the previously developed concepts are applied to instantiate autonomic cooperative deployments of Fixed Relay Nodes chosen out of a mesh of Fixed Radio Access Points. Finally, the work is extended to emergency networks by investigating the preferred network configuration strategies, enhanced with cooperative behaviours expressed by autonomic cooperative nodes, through the use of relaying.

Following the introductory part contained in Chapter 1, the book is organised as explained below. In particular, the idea of spatio-temporal processing is outlined in Chapter 2 and it is complemented by the analysis of conventional and cooperative relaying in Chapter 3. Then Chapter 4 follows, where routing information enhanced cooperative transmission is introduced, to be incorporated into autonomic cooperative system design in Chapter 5. Next, the previously analysed concepts are extended to cooperative autonomic network deployments in Chapter 6, and, eventually, they are enhanced with the aspects of autonomic emergency communications in Chapter 7. The book is concluded in Chapter 8.

Chapter 2
Spatio-Temporal Processing

2.1 Introduction

Spatio-temporal processing emerged as one of the key achievements towards the provision of high data rate, reliable wireless communications. Most recently, the rationale behind this concept has been mapped onto networked systems under the name of cooperative transmission. This chapter provides a general background on spatio-temporal processing to form the basis of further investigations outlined in the remainder of this book. In particular, the gains achievable in Multiple Input Multiple Output Channels are first quantified. Then, the relevant diversity techniques are discussed together with the role of diversity order and diversity gain. Following, Space-Time Block Coding and Space-Time Trellis Coding techniques are introduced and supported with performance results. Eventually, the context of Layered Space-Time Coding is provided. This analysis is further complemented in the next chapter with the definition of Equivalent Distributed Space-Time Block Encoder.

2.2 Multiple Input Multiple Output Channel

Traditionally, before the emergence of systems capable of exploiting the transmit diversity over spatial dimension, signals were generally transmitted in time and frequency domains [25]. This was performed either with the aid of Single Input Single Output (SISO) or Single Input Multiple Output (SIMO) radio channels. Afterwards, when it turned out that additional information may be equally well conveyed using spatial diversity, global research started focusing on Multiple Input Single Output (MISO) and Multiple Input Multiple Output (MIMO) technologies. The latter forms the most general approach with the former being, in fact, its special case. Soon after, the MIMO technology was mapped onto cooperative networked systems [2], as it will be explained in the following chapters [3], [26]. The wireless MIMO channel (Figure 2.1) is usually defined with the aid of a channel matrix $H_{N \times M}$ (2.1),

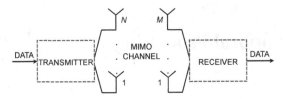

Fig. 2.1 Multiple Input Multiple Output System

containing the coefficients $h_{i,j}$ referring to the radio links between each transmitting antenna i $(1 \leq i \leq N)$ and each receiving antenna j $(1 \leq j \leq M)$.

$$H_{N \times M} = \begin{bmatrix} h_{1,1} & h_{1,2} & \cdots & h_{1,M} \\ h_{2,1} & h_{2,2} & \cdots & h_{2,M} \\ \vdots & \vdots & \ddots & \vdots \\ h_{N,1} & h_{N,2} & \cdots & h_{N,M} \end{bmatrix} \quad (2.1)$$

As it was shown in [15] for an SISO system featuring Additive White Gaussian Noise (AWGN) channel, the data rate boundary for a single user can be expressed as (2.2), where $|h|^2$ represents the channel gain. This equation combines the channel capacity C with channel bandwidth B and the Signal-to-Noise Ratio (SNR). The latter is expressed as the quotient of the transmitted signal power P_T to the noise power σ^2. It might seem then that the easiest way to increase the capacity, and thus the attainable data rate, would be to widen the bandwidth. Unfortunately, at least from the commercial perspective, taking into account that the bandwidth is scarce and deficit component, such an approach would render it too costly, inefficient, and, consequently, hardly acceptable.

$$C = B \log_2 \left(1 + \frac{P_T}{\sigma^2} |h|^2 \right) \quad (2.2)$$

Another potential approach would be to increase the SNR. Unfortunately, it would require an increase in the power of the transmitted signal, and, consequently, it would enlarge the co-channel and inter-channel interference levels. Moreover, due to the logarithmic relation between both parameters, the effective channel capacity gain would be less significant compared to the first case.

The most relevant way of addressing this issue is then to employ the MIMO processing [20]. Generally, if the number of both transmitting antennas N and receiving antennas M antennas is equal to 1, it is possible to gain merely about 1 bit/Hz for the increase in SNR of 3 dB [5]. However, if Multi-Element Arrays (MEAs) are employed at both sides of the wireless link, and they are of the same size equal to N, the capacity may scale linearly with N [11] and, consequently, it becomes feasible to achieve almost N bits per Hz [5]. This phenomenon may be very neatly explained with the aid of the Singular Value Decomposition (SVD) theorem, as described in [22]. In particular, the channel matrix $H_{N \times M}$ can be written in the following way (2.3):

$$H = UDV^H \tag{2.3}$$

where D is a non-negative and diagonal matrix of size $M \times N$, U and V are unitary matrices of size $M \times M$ and $N \times N$, respectively, and the upper index ^{H} denotes the Hermitian transpose. It means that $UU^H = I_M$ and $VV^H = I_N$, where I_M is an identity matrix of size $M \times M$, and I_N is an identity matrix of size $N \times N$. The diagonal entries of D are then non-negative square roots of the eigenvalues of the matrix HH^H, denoted by λ, and, defined as (2.4) [22]:

$$HH^H y = \lambda y, \qquad y \neq 0 \tag{2.4}$$

where y is an eigenvector of size $M \times 1$, associated with λ. For the remaining details the reader is referred to [22]. What is necessary for the further analysis carried out in this book is that one may now think of an equivalent MIMO channel comprising k uncoupled parallel sub-channels, where k is the rank of the channel matrix H, and, at most, is equal to m, i.e. the minimum value of both N and M. Consequently, the channel capacity formula can be expressed as (2.5) [22]:

$$C = B \log_2 \det \left[I_m + \frac{P_T}{N\sigma^2} Q \right] \tag{2.5}$$

where Q equals to HH^H for $N < M$ and to $H^H H$ for $N > M$, respectively.

Now, taking into account the case where $M = N$, and assuming that transmitting and receiving antennas are connected exclusively by the aforementioned orthogonal parallel sub-channels, H may be written as (2.6) [22]:

$$H = \sqrt{N} I_N \tag{2.6}$$

where $\sqrt{N}$ is a scaling factor pertaining to power normalisation. Finally, after including (2.6) in (2.5), the obtained capacity C is equal to (2.7):

$$C = NB \log_2 \left(1 + \frac{P_T}{\sigma^2} \right) \tag{2.7}$$

This indeed shows clearly that it is possible to achieve extremely high throughputs in MIMO systems. In general, two approaches are possible [13]. On the one hand, one can create a highly effective diversity scheme for the purposes of increasing the robustness of the system against the impairments induced by the wireless radio channel [24]. On the other hand, one can transmit multiple parallel data streams instead, and therefore increase the system throughput.

2.3 Diversity Techniques

Before the idea of space-time coding has been introduced, the relevant diversity techniques are first briefly characterised [23], [22], [8]. These techniques are common means of combating the effects of multipath fading, as well as improving the transmission reliability [25]. The diversity phenomenon is based on the assumption that there are multiple replicas of the transmitted signal available at the receiver. Each of them conveys the same information but the fading, they are subject to, is usually almost uncorrelated. Consequently, it is very unlikely that all the replicas might encounter a deep fade simultaneously, and, hence, the probability of proper reception increases. The most general classification mentions diversity in time, frequency and space domains.

Time diversity is also known as temporal diversity [8], and it assumes the transmission of multiple replicas of the signal in different time slots. The required separation between these slots must be at least equal to the coherence time of the radio channel [22]. The coherence time is defined as the time during which the autocorrelation function of the channel impulse response is approximately non-zero [6]. Such an approach to diversity results in decoding delays and it is most suitable for fast fading environments where the coherence time is short. Frequency diversity, in turn, exploits different frequencies for the purposes of transmitting the replicas of the original signal. Obviously, these frequencies must be appropriately separated to ensure that different parts of the spectrum will be subjected to independent fades [8]. Such a separation is determined by the coherence bandwidth defined as the frequency range across which the entire signal bandwidth is highly correlated. In other words, it means that fading is roughly equal over this range [6]. Consequently, if the fading statistics for different frequencies are supposed to be essentially uncorrelated, the frequency separation of the order of several times the channel coherence bandwidth is necessary [22]. Space diversity, unlike the other two techniques, induces no loss in bandwidth efficiency [22]. In this case, multiple antennas are used, which must be separated by a few wavelengths[1] to guarantee that the replicas of the transmitted signal are uncorrelated. There are two examples of space diversity: polarisation diversity and angle diversity [22], [8]. In the first case, signals of horizontal and vertical polarisation are transmitted and received by two sets of differently polarised antennas. This is to ensure that there would be no correlation between the two signals, even if the antennas were not separated by a few wavelengths. The angle diversity, in turn, is applicable to carrier frequencies larger than 10 GHz. Such environments are characterised by rich scattering in the space domain and, therefore, it suffices to use two highly directional receiving antennas, pointed at two different directions, to fully gain from this type of diversity.

Given the scope of this book, in the following, more attention is paid to the categorisation of spatial diversity. One should note that depending on whether multiple antennas, separated spatially, are located at the transmitter or at the receiver, two subcategories can be distinguished: reception diversity and transmission diversity.

---

[1] In [8] this type of separation is also referred to as the coherence distance.

One of the least sophisticated approaches to reception diversity is selection com-

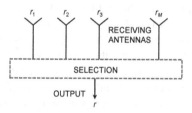

Fig. 2.2 Selection combining

bining [22], as depicted in Figure 2.2. In this method this signal r_i, $(i = 1,\ldots,M)$ is chosen which is characterised by the highest value of the instantaneous SNR. For this purpose, all the diversity branches would need to be monitored continuously and simultaneously. Therefore, a suboptimal solution is also known which is referred to as switched combining or scanning diversity. Here, the reduced complexity is traded off against the lower performance. In other words, this diversity branch remains selected which is able to maintain the SNR above a specified threshold. Following, there is the commonly known Maximal Ratio Combining (MRC) solu-

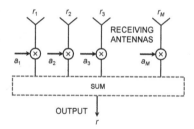

Fig. 2.3 Maximal ratio combining

tion, as described in Figure 2.3. It is a linear method, where signals coming from distinct diversity branches are first weighted with the use of a_i coefficients, and, then, added together. Each coefficient a_i is defined as (2.8):

$$a_i = A_i e^{-j\varphi_i} \tag{2.8}$$

where A_i is the amplitude and φ_i is phase of the signal r_i received by the receiving antenna i. The total received signal r can be then obviously expressed as (2.9):

$$r = \sum_{i=1}^{M} a_i r_i \tag{2.9}$$

MRC is an optimum combining method in the sense that it maximises the output SNR [22]. Finally, there is also a suboptimal version of the MRC known, which does not require estimation of the fading amplitude for each of the diversity branches. Instead, it assumes the amplitudes A_i are all equal to 1, and, hence, it is known under the name of Equal Gain Combining (EGC). The performance of EGC is only slightly worse compared to MRC and the implementation complexity is significantly reduced [22]. In general, the implementation of reception diversity is considered at the Base Station (BS), because it might be very difficult to equip User Terminals (UTs) in the form of mobile phones with multiple antennas and provide batteries that would be capacious enough to survive the existence of separate Radio Frequency (RF) chain for each of these antennas.

This is why, in general, the focus is on the transmit diversity, even though it is perceived more difficult to exploit for some reasons [22]. Firstly, once the signals transmitted from multiple antennas arrive at the receiver, they are spatially mixed. Therefore, additional processing both at the transmitter and the receiver is definitely required. Secondly, unless it is fed back from the receiver, the transmitter does not have instantaneous information about the channel parameters. This is in stark contrast to receive diversity, where the receiver is usually able to estimate the channel coefficients. There are a number of schemes and the delay transmit diversity will be presented here as the classic example [22], [9]. In this case, the copies of the transmitted signal are delayed according to the scheme presented in Figure 2.4. As

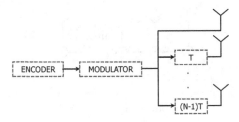

Fig. 2.4 Delay transmit diversity

a result, the receiver observes the original copy of the signal, as if it was distorted by a multipath propagation and can use a Maximum Likelihood Sequence Estimator (MLSE) or Minimum Mean Square Error (MMSE) equaliser to obtain diversity gain [22]. In comparison, Space-Time Coding (STC) combines both space diversity and temporal diversity.

Analysing diversity methods, one also needs to define the so called diversity order and diversity gain. In general, the higher the number of independent fading branches, paths or receiving antennas in SIMO channels, the higher the diversity order, and thus the better performance of such a system [8]. In particular, assuming maximum-likelihood detection or maximum-ratio combining, the average probability of error for high SNR values can be written as (2.10) [9]:

$$P(e) \sim G_c (SNR)^{-G_d} \qquad (2.10)$$

where G_c denotes the coding gain[2] provided by block or convolutional coding in the time domain, whereas G_d is the aforementioned diversity order. If the $P(e)$ curve is plotted as a function of SNR on a log-log scale, then G_c determines the horizontal position of this curve and G_d corresponds to its slope [13]. Thanks to the diversity order, the diversity gain is observable which is defined as the gain provided by spatial diversity across channels either at the transmitter, receiver, or both of them [8]. Generally, complete Channel State Information (CSI) is required at the transmitter when transmission diversity techniques are employed [8]. However, space-time block coding, to be presented in the following section, does not require CSI at the transmitter [1]. What is more, it is characterised by full diversity order equal to the product of the number of transmitting and receiving antennas [8].

2.4 Space-Time Block Coding

There are a few spatio-temporal processing techniques which can be employed for the purposes of pre-processing the transmitted signals in such a way that they are more robust to the impairments induced by wireless radio propagation [25]. Among them there is space-time block coding introduced by Alamouti in [1] which offers diversity gain but no coding gain. For this reason, despite its name, space-time block coding also happens to be perceived as a modulation technique rather than a coding technique. In particular, it was designed to provide additional spatio-temporal diversity in wireless systems for the purposes of enhancing transmission reliability. As already mentioned, when compared to the classic solutions based on reception diversity, space-time block coding allows to shift the complexity connected with multiple antennas from small mobile UTs to BSs. Among the most significant advantages of this approach is the reduced complexity of UTs, lower cost of installing one MEA at the BS only, as well as the possibility of guaranteeing reasonable spacing among elements of such an antenna array. The base G_2 space-time block code is defined as follows (2.11):

$$G_2 = \begin{bmatrix} x_1 & x_2 \\ -x_2^* & x_1^* \end{bmatrix} \tag{2.11}$$

This code may be used in a system employing two transmitting and any number of receiving antennas. More specifically, in the first time slot, the x_1 and x_2 symbols are sent by the first and second transmitting antenna, respectively, and then, in the second time slot, the $-x_2^*$ and x_1^* symbols are transmitted alike. For further details, the reader is referred to [1]. Besides, also other space-time block codes are known [17], [16], such as e.g. G_3 (2.12), G_4 (2.13), H_3 (2.14) and H_4 (2.15). These codes may be especially applicable to antenna arrays of greater sizes. One should also note that there is a trade-off between the robustness of each of these codes and their rate R,

---

[2] In [9] coding gain is also referred to as coding advantage.

which is strictly connected with the number of transmitting antennas. In fact, the code rate is equal to 1 in the case of the G_2 code only.

$$
G_3 = \begin{bmatrix}
x_1 & x_2 & x_3 \\
-x_2 & x_1 & -x_4 \\
-x_3 & x_4 & x_1 \\
-x_4 & -x_3 & x_2 \\
x_1^* & x_2^* & x_3^* \\
-x_2^* & x_1^* & -x_4^* \\
-x_3^* & x_4^* & x_1^* \\
-x_4^* & -x_3^* & x_2^*
\end{bmatrix}
\tag{2.12}
$$

$$
G_4 = \begin{bmatrix}
x_1 & x_2 & x_3 & x_4 \\
-x_2 & x_1 & -x_4 & x_3 \\
-x_3 & x_4 & x_1 & -x_2 \\
-x_4 & -x_3 & x_2 & x_1 \\
x_1^* & x_2^* & x_3^* & x_4^* \\
-x_2^* & x_1^* & -x_4^* & x_3^* \\
-x_3^* & x_4^* & x_1^* & -x_2^* \\
-x_4^* & -x_3^* & x_2^* & x_1^*
\end{bmatrix}
\tag{2.13}
$$

$$
H_3 = \begin{bmatrix}
x_1 & x_2 & \frac{x_3}{\sqrt{2}} \\
-x_2^* & x_1^* & \frac{x_3}{\sqrt{2}} \\
\frac{x_3^*}{\sqrt{2}} & \frac{x_3^*}{\sqrt{2}} & \frac{\left(-x_1-x_1^*+x_2-x_2^*\right)}{\sqrt{2}} \\
\frac{x_3^*}{\sqrt{2}} & -\frac{x_3^*}{\sqrt{2}} & \frac{\left(x_2+x_2^*+x_1-x_1^*\right)}{\sqrt{2}}
\end{bmatrix}
\tag{2.14}
$$

$$
H_4 = \begin{bmatrix}
x_1 & x_2 & \frac{x_3}{\sqrt{2}} & \frac{x_3}{\sqrt{2}} \\
-x_2^* & x_1^* & \frac{x_3}{\sqrt{2}} & -\frac{x_3}{\sqrt{2}} \\
\frac{x_3^*}{\sqrt{2}} & \frac{x_3^*}{\sqrt{2}} & \frac{\left(-x_1-x_1^*+x_2-x_2^*\right)}{\sqrt{2}} & \frac{\left(-x_2-x_2^*+x_1-x_1^*\right)}{\sqrt{2}} \\
\frac{x_3^*}{\sqrt{2}} & -\frac{x_3^*}{\sqrt{2}} & \frac{\left(x_2+x_2^*+x_1-x_1^*\right)}{\sqrt{2}} & -\frac{\left(x_1+x_1^*+x_2-x_2^*\right)}{\sqrt{2}}
\end{bmatrix}
\tag{2.15}
$$

Although more reliable, the other codes offer worse rates. In particular, the H_3 and H_4 codes are characterised by rate equal to $\frac{3}{4}$, whereas G_3 and G_4 achieve the rate equal to $\frac{1}{2}$.

Looking at the process of reception, space-time block decoder could operate solely with the use of a single receiving antenna. However, for the best performance, it is strongly recommended to use a larger receiving antenna array. The signal received by a receiving antenna j may be written as (2.16):

$$
r_t^j = \sum_{i=1}^{N} h_{i,j} s_t^i + \eta_t^j
\tag{2.16}
$$

where $h_{i,j}$ denotes the channel coefficient (see MIMO channel matrix 2.1), s_t^i represents the symbol transmitted by antenna i and the noise samples η_t^j are modelled by the complex Gaussian process with zero mean and $N_0/2$ variance per dimension. The main feature of space-time block codes, being also the main condition under which the operation of decoding may be successfully performed, is their orthogonality [1], [16], [10]. This condition is defined as (2.17):

$$G_N G_N^H = \left(\sum_{i=1}^{N} |x_i|^2 \right) I_N \tag{2.17}$$

where N is equal to the number of transmitting antennas and I_N is an identity matrix of size $N \times N$. The process of decoding is based on a maximum-likelihood detection aiming to minimise the decision metric given by the formula (2.18) [16], which can be easily derived on the basis of the theory provided, for example, in [6].

$$z = \sum_{t=1}^{L} \sum_{j=1}^{M} \left| r_t^j - \sum_{i=1}^{N} h_{i,j} s_t^i \right|^2 \tag{2.18}$$

It means that, for a given code, these potentially transmitted symbols are chosen, which minimise this metric. In this section, the validation results of three different systems featuring the AWGN MIMO channel are provided. The power emitted by each of the transmitting antennas is always normalised so that the total transmitted power is guaranteed to be equal to 1. The SNR at the receiving antenna j is then defined as the total received signal power to the noise power ratio. Each time 10 million bits are transmitted and up to 4 receiving antennas are used. In Figure 2.5, Figure 2.6 and Figure 2.7, the results showing performance of G_2, G_3 and H_3 under the above conditions and for different numbers of receiving antennas are presented, respectively [25].

2.5 Space-Time Trellis Coding

Space-time trellis coding was introduced by Tarokh[3] in [18] and then further investigated in [19]. In contrast to space-time block coding, mostly perceived as a modulation technique (see Section 2.4), space-time trellis coding aims to introduce additional relations among specific sequences transmitted by distinct antennas, as well as the symbols constituting these sequences. As a result, apart from diversity gain, additional coding gain may be observed. The base space-time trellis code proposed in [18], which exploits the Phase Shift Keying (PSK) modulation scheme (4-PSK) is presented in Figure 2.8 [25]. The numbers placed to the left of the trellis diagram should be interpreted in the following way: the most significant digit represents the current state, whereas the least significant one corresponds to the in-

---

[3] However, one should also note that besides Alamouti, Tarokh and Poon investigated a concatenation of a space-time block encoder with an outer trellis code in [12].

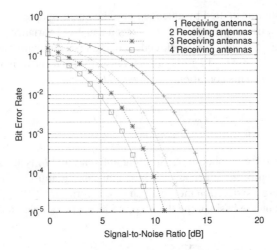

Fig. 2.5 Performance of G_2 code for 1, 2, 3 and 4 receiving antennas

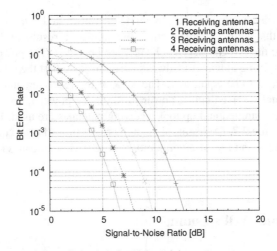

Fig. 2.6 Performance of G_3 code for 1, 2, 3 and 4 receiving antennas

put and, therefore, also to the next state. It means that the consecutive pairs of the encoder input bits determine the transition from the current state to the following one. In other words, two symbols are relayed to transmitting antennas, so the first antenna transmits the channel symbol informing about the current state, while the second antenna transmits the channel symbol informing about the next state.

The process of decoding is based on the well-known Viterbi algorithm [21] and each transition on the trellis is assigned a metric, which is calculated according to the formula (2.19) [18]:

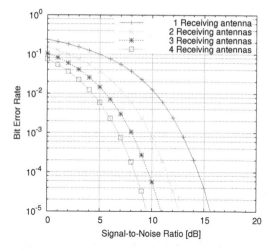

Fig. 2.7 Performance of H_3 code for 1, 2, 3 and 4 receiving antennas

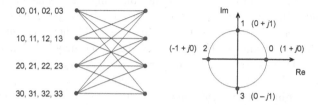

Fig. 2.8 Basic space-time trellis code exploiting 4-PSK modulation

$$w_{x,y} = \sum_{j=1}^{M} \left| r_t^j - \sum_{i=1}^{N} h_{i,j} s_t^i \right|^2 \qquad (2.19)$$

where x and y denote the states, being the beginning and the end of a given transition. In case there are no channel impairments, the decoding procedure for an example input sequence $\{1,0,3,1,2,2\}$ and the space-time trellis code from Figure 2.8 would be carried out as shown in Figure 2.9. Let us assume that initially the encoder remains in the zero state, and at each moment t, it is possible to move from the current state to the next one, in the subsequent moment $t+1$, under one of the input values $\{0,1,2,3\}$. That is why, the example input sequence will result in relaying the following sequence of symbol pairs $\{01,10,03,31,12,22\}$ to transmitting antennas. Therefore, the first antenna will transmit signals corresponding to the symbol sequence $\{0,1,0,3,1,2\}$, and at the same time, the second antenna will transmit signals corresponding to the symbol sequence $\{1,0,3,1,2,2\}$. It means that the following modulated sequences $\{1,j,1,-j,j,-1\}$ and $\{j,1,-j,j,-1,-1\}$ will be observed at the output of the modulator.

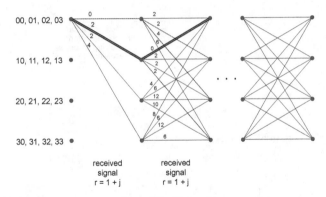

Fig. 2.9 STTC decoding process according to Viterbi algorithm

In particular, in the first modulation interval, the (s_0, s_1) pair, equal to $(1 + j0, 0 + j1)$ is transmitted, and the received signal (defined as 2.16) may be written as $r_1^1 = 1 + j$. Consequently, according to (2.19), the following metrics are calculated for the corresponding transitions: $w_{0,0} = 2, w_{0,1} = 0, w_{0,2} = 4, w_{0,3} = 2$. Next, the same procedure is performed in the second modulation interval, where the pair $(s_0, s_1) = (0 + j1, 1 + j0)$ is transmitted. Here, the received signal may be written as $r_2^1 = j + 1$ and so the metrics are: $w_{0,0} = 2, w_{0,1} = 0, w_{0,2} = 2, w_{0,3} = 4, w_{1,0} = 0, w_{1,1} = 2, w_{1,2} = 2, w_{1,3} = 2, w_{2,0} = 4, w_{2,1} = 8, w_{2,2} = 10, w_{2,3} = 2, w_{3,0} = 2, w_{3,1} = 0, w_{3,2} = 4, w_{3,3} = 2$. According to the Viterbi algorithm, in case there are a number of paths ending up in the same state, the one characterised by the lowest cumulative metric is chosen. In the presented example, there are four candidate paths selected. Therefore, if the decision were to be made at this stage, the path visible as double line would be finally picked. However, one should note that usually the trellis should be as deep as 3 to 5 times the constraint length of the convolutional code in order to make the decisions reliable [23]. There were also other space-time trellis codes proposed in [18] and two additional examples are depicted in Figure 2.10 and Figure 2.11.

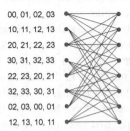

Fig. 2.10 Example space-time trellis code for 4-PSK modulation

Similarly to the simulation assumptions made in Section 2.4, up to 4 receiving antennas are utilised and each time 10 million bits are transmitted over AWGN chan-

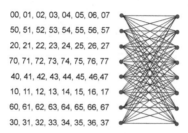

00, 01, 02, 03, 04, 05, 06, 07
50, 51, 52, 53, 54, 55, 56, 57
20, 21, 22, 23, 24, 25, 26, 27
70, 71, 72, 73, 74, 75, 76, 77
40, 41, 42, 43, 44, 45, 46,47
10, 11, 12, 13, 14, 15, 16, 17
60, 61, 62, 63, 64, 65, 66, 67
30, 31, 32, 33, 34, 35, 36, 37

Fig. 2.11 Example space-time trellis code for 8-PSK modulation

nel. The power emitted by each of the transmitting antennas is always normalised and, as a result, the total transmitted power is equal to 1. The SNR at the receiving antenna j is defined as a ratio between the cumulative received signal power and the noise power. The obtained results are depicted in Figure 2.12 [25].

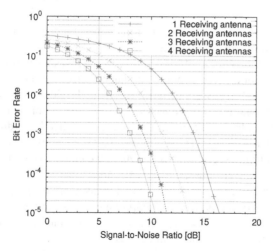

Fig. 2.12 Performance of STTC code for 1, 2, 3 and 4 receiving antennas

2.6 Layered Space-Time Coding

By contrast with Space-Time Block Coding and Space-Time Trellis Coding, in the classic Layered Space-Time Coding (LSTC) case, proposed in [4], N information streams are transmitted simultaneously over the same frequency band with the use of N transmitting antennas [22]. The simplest, uncoded Layered Space-Time Architecture, which is commonly referred to as Vertical Layered Space-Time or Vertical Bell Laboratories Layered Space-Time system, is outlined in Figure 2.13. Here, the

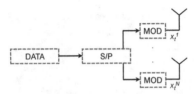

Fig. 2.13 Vertical layered space-time architecture

input data stream is demultiplexed into N sub-streams, which are then modulated separately and transmitted. Each processing chain is referred to as a separate layer which explains the name of this approach. An example of a transmission matrix X for a system employing three transmitting antennas is given by (2.20):

$$X = \begin{bmatrix} x_1^1 & x_2^1 & x_3^1 & x_4^1 & \cdots \\ x_1^2 & x_2^2 & x_3^2 & x_4^2 & \cdots \\ x_1^3 & x_2^3 & x_3^3 & x_4^3 & \cdots \end{bmatrix} \tag{2.20}$$

It means that the entry x_t^i is transmitted from antenna i at time t. Additional improvements to this approach are outlined, e.g. in [7].

The receiver, in turn, exploits $M = N$ receiving antennas for the purposes of performing the process of detection comprising interference suppression and cancellation. The distinguished signals are then decoded with the aid of the relevant conventional, one-dimensional block or convolutional decoding algorithm, which results in a much lower complexity as compared to Space-Time Trellis Coding (STTC). In general, the structure of the LSTC system can be also perceived as a synchronous Code Division Multiple Access (CDMA), where the number of transmitting antennas is equal to the number of users [22], [14]. More sophisticated approaches were also proposed, where the conventional, one-dimensional[4] block or convolutional codes are exploited for the purposes of improving the performance of the system. These include the Horizontal Layered Space-Time, Diagonal Layered Space-Time or Threaded Layered Space-Time architectures.

2.7 Conclusion

In this chapter, the most important aspects of spatio-temporal processing were highlighted to pave the ground for the investigations to be carried out in the following chapters. In particular, the MIMO channel was introduced and the reasons for the achievable gain in throughput were explained. Then the rationale behind the relevant diversity techniques was discussed before both the Space-Time Block Coding and Space-Time Trellis Coding techniques were introduced and contrasted with Layered Space-Time Coding. Especially the notion of STBC will be further extended to be

---

[4] In the space domain [22].

mapped onto the Cooperative Relaying case in the following chapter, where these concepts are applied to networked systems operating in a distributed manner.

References

1. S. Alamouti. A Simple Transmit Diversity Technique for Wireless Communications. *IEEE Journal on Selected Areas in Communications*, 16(8):1451–1458, Oct. 1998.
2. M. Dohler and Y. Li. *Cooperative Communications - Hardware, Channel & PHY*. Wiley, 2010.
3. K. Doppler, S. Redana, M. Wódczak, P. Rost, and R. Wichman. Dynamic resource assignment and cooperative relaying in cellular networks: Concept and performance assessment. *EURASIP Journal on Wireless Communications and Networking*, Jul. 2007.
4. G. J. Foschini. Layered Space-Time Architecture for Wireless Communications in a Fading Environment When Using Multi-Element Antennas. *Bell Labs Technical Journal*, 1(2):41–59, 1996.
5. G. J. Foschini and M. J. Gans. On Limits of Wireless Communications in Fading Environment when Using Multiple Antennas. *Wireless Personal Communications*, 6:311–335, 1998.
6. A. Goldsmith. *Wireless Communications*. Cambridge University Press, 2005.
7. B. Hassibi and B. M. Hochwald. High-rate codes that are linear in space and time. *IEEE Transactions on Information Theory*, 48(7):1473–1484, Jun. 2002.
8. M. Jankiraman. *Space-Time Codes and MIMO Systems*. Artech House, 2004.
9. E. G. Larsson and P. Stoica. *Space-Time Block Coding for Wireless Communications*. Cambridge University Press, 2003.
10. E.G. Larsson and P Stoica. Mean square error optimality of orthogonal space-time block codes. *IEEE International Conference on Communications, ICC*, pages 2272–2275, May 2003.
11. A. Lozano, F. R. Farrokhi, and R. A. Valenzuela. Lifting the limits on high-speed wireless data access using antenna arrays. *IEEE Communications Magazine*, 39(9):156–162, Sep. 2001.
12. S. M.Alamouti, V.Tarokh, and P.Poon. Trellis coded modulation and transmit diversity: Design criteria and performance evaluation. *IEEE International Conference on Universal Personal Communications, ICUPC*, pages 703–707, Oct. 1998.
13. A.F. Molisch and M.Z. Win. MIMO Systems with Antenna Selection. *IEEE Microwave Magazine*, 5(4):46–56, Mar. 2004.
14. S. Sfar, R.D. Murch, and K.B. Letaief. Layered space-time multiuser detection over wireless uplink systems. *IEEE Transactions on Communications*, 2(4):653–668, Jul. 2003.
15. C. E. Shannon. A Mathematical Theory of Communication. *Bell System Technical Journal*, 27:379–423 and 623–656, Jul. and Oct. 1948.
16. V. Tarokh, H. Jafarkhani, and A. R. Calderbank. Space-time block codes from orthogonal designs. *IEEE Transactions on Information Theory*, 45(5):1456–1467, Jul. 1999.
17. V. Tarokh, H. Jafarkhani, and A. R. Calderbank. Space-time block coding for wireless communications: performance results. *IEEE Journal on Selected Areas in Communications*, 17(3):451–460, Mar. 1999.
18. V. Tarokh, N. Seshadri, and A. R. Calderbank. Space-Time Codes for High Data Rate Wireless Communication: Performance Criterion and Code Construction. *IEEE Transactions on Information Theory*, 44(2):744–765, Mar. 1998.
19. V. Tarokh, N. Seshadri, and A. R. Calderbank. Space-Time Codes for High Data Rate Wireless Communication: Performance Criteria in the Presence of Channel Estimation Errors, Mobility, and Multiple Paths. *IEEE Transactions on Communications*, 47(2):199–207, Feb. 1999.
20. I. E. Telatar. Capacity of multi-antenna Gaussian channels. *European Transactions on Telecommunications*, 10(6):585–595, Nov.-Dec. 1999.
21. A. J Viterbi. Error Bounds for Convolutional Codes and an Asymptotically Optimum Decoding Algorithm. *IEEE Transactions on Information Theory*, 13(2):260–269, Apr. 1967.

22. B. Vucetic and J. Yuan. *Space-Time Coding*. John Wiley & Sons, 2003.
23. K. Wesołowski. *Mobile communications systems*. John Wiley and Sons, 2002.
24. M. Wódczak. On the Adaptive Approach to Antenna Selection and Space-Time Coding in Context of the Relay Based Mobile Ad-hoc Networks. *XI National Symposium of Radio Science URSI, Poznań, Poland*, pages 138–142, Apr. 2005.
25. M. Wódczak. *On Routing information Enhanced Algorithm for space-time coded Cooperative Transmission in wireless mobile networks*. PhD thesis, Faculty of Electrical Engineering, Institute of Electronics and Telecommunications, Poznań University of Technology, Poland, Sep. 2006.
26. M. Wódczak. Autonomic Cooperative Networking for Wireless Green sensor Systems. *International Journal of Sensor Networks (IJSNet)*, 10(1/2), 2011.

Chapter 3
Conventional and Cooperative Relaying

3.1 Introduction

Relaying techniques are becoming more and more widespread across wireless communications, involving ad hoc and mesh networks, as well as cellular systems. This chapter provides a comprehensive, where necessary, and concise, overall, description of both the conventional and cooperative relaying protocols. Initially, a general classification of relaying techniques is provided, and then, an adaptive conventional scenario of Manhattan type is analysed. Following, based on the investigations provided in the previous chapter, the main emphasis is put on Distributed Space-Time Block Coding, presented as a special case of Virtual Antenna Arrays. As a result, the Equivalent Distributed Space-Time Block Encoder is defined, which will form the basis for further investigations related to enhancing cooperative transmission with additional routing information, as described in the following chapter, as well as for evaluating relevant cooperative relaying scenarios later on.

3.2 Classification of Relaying Protocols

The method of conventional relaying, as depicted in Figure 3.1, is also known under the name of Layer 3 Decode-and-Forward (L3DF) scheme [26] and it consists of two phases. First, the Source Node (SN) sends its information to the Relay Node (RN). The RN fully decodes the received signal, then encodes it again and sends to the Destination Node (DN). Such an approach makes it feasible either to reduce the transmitted power or extend the range, however, it cannot offer diversity. It is referred to as L3DF because, taking into account the Open Systems Interconnection (OSI) layered model, the operation of relaying is performed at the network layer [13].

Cooperative relaying, instead, is based on cooperation among intermediary nodes, where some of them act as RNs to assist the process of transmission between the SN

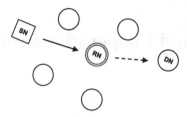

Fig. 3.1 Conventional relaying

and DN [7], [23]. An example is depicted in Figure 3.2, where the process of transmission is also carried out in two phases. First, both the DN and RN receive the transmitted signal and then the RN may additionally resend its copy towards the destination in order to, potentially, improve the performance of such a system by providing extra diversity. There are a number of different approaches to cooperative

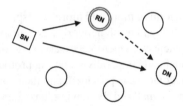

Fig. 3.2 Cooperative relaying

relaying known and this idea is also referred to as cooperation diversity, cooperative diversity, virtual antenna arrays or coded cooperation [26]. In this book, however, the term cooperative transmission is preferred [22]. Moreover, on the basis of the work described in [15] and according to the later classification given in [25] and then extended in [26] and [13], the cooperative relaying protocols can be categorised with regard to either the forwarding strategy or the protocol nature. In the first case, one may distinguish the Amplify-and-Forward, Decode-and-Forward and Decode-and-Reencode categories. The Amplify-and-Forward protocols are also referred to as non-regenerative ones, while the Decode-and-Forward and Decode-and-Reencode ones belong to the regenerative category [13]. The Decode-and-Reencode group additionally includes fixed, adaptive and feedback protocols. All of these approaches are briefly characterised in Table 3.1 and Table 3.2.

Moreover, a general cooperative scheme was presented in [12] (Figure 3.3), where the following notation was proposed (3.1):

$$p = \{M, m_{Tx}, m_{rel}, m_{Rx}\} \qquad (3.1)$$

Specifically, p denotes a set of parameters containing the number of relay nodes M, as well as the number of antennas at the SN, RN and DN, defined by m_{Tx}, m_{rel} and

Table 3.1 Cooperative protocols classification according to forwarding strategy

Protocol	Description
Amplify-and-Forward	Before retransmission takes place, the RN acts as an analogue repeater and only amplifies the received signal causing noise enhancement in the relay path.
Decode-and-Forward	Before retransmission takes place, the RN attempts to fully decode, regenerate and reencode the received signal, potentially causing propagation of decoding errors, leading to wrong decisions at the destination.
Decode-and-Reencode	Before retransmission takes place, the RN fully decodes, regenerates and constructs a new code word, different from the source one, which may possibly also result in propagation of errors, but enables parallel channel coding.

Table 3.2 Cooperative protocols classification according to protocol nature

Protocol	Description
Fixed protocol	The RN always, possibly also after having performed some processing, forwards the received signal.
Adaptive protocol	The RN autonomically decides whether to forward the received signal or not.
Feedback protocol	The RN assists in the transmission only if it receives an explicit request from the destination.

m_{Rx}, respectively. For example, an SISO system featuring an additional relay node is described as $p = \{1,1,1,1\}$, while an MISO system, where the SN is equipped with 2 transmit antennas, is described as $p = \{1,2,1,1\}$. In the following sections, first

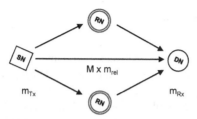

Fig. 3.3 General scheme of cooperation among relay nodes

certain aspects of conventional relaying will be outlined, and then, further details about more complex cooperative schemes will be provided as a foundation for the remaining chapters.

3.3 Adaptive Conventional Relaying

Conventional relaying is presented with the use of a fixed deployment concept also known as the Manhattan scenario (Figure 3.4). This scenario consists of one BS

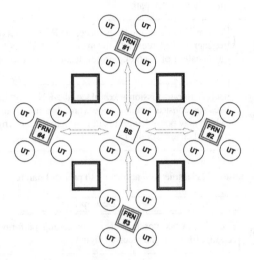

Fig. 3.4 Fixed deployment concept of Manhattan type

situated in the centre, four Fixed Relay Nodes (FRNs) separated by the buildings, and a number of UTs around each of these [8], [9]. While different approaches to frame structuring are possible, the one presented in Figure 3.5 is assumed here. In

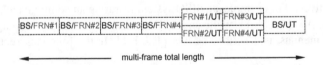

Fig. 3.5 Case 2 multi-frame structure

this case the spatial independence between two pairs of FRNs is guaranteed [18], as the multi-frame is subdivided into three phases. During the first phase, all the four FRNs are served by the BS one by one and then, in the second phase, the two spatially independent FRN couples serve their UTs. Finally, the BS serves its UT in the last, third phase. This approach can be further optimised by taking into account the traffic offered by distinct nodes and the knowledge about buffer utilisation [22], [10].

In such a system, where one may gain mostly from the network layer optimisation, different levels of Quality of Service (QoS) could be supported, depending on the requirements of the specific users and their applications. However, to enable QoS provision from a wider perspective, there must be a resource management mechanism active at the BS which would be capable of allocating specific service times. In particular, either the BS or any of the FRNs are equipped with buffer memory where they store packets. Such memory is limited and if the number of packets lost is to be minimised, the system must be sensitive to any changes in the traffic offered by distinct nodes. This problem can be dealt effectively with the aid of a scheduler. While there are different scheduling algorithms known, the Weighted Round-Robin scheduler appears to be the most flexible solution. This is because the users are allowed to utilise the bandwidth with respect to their weights [17]. The described throughput

---

Algorithm 1 Throughput maximisation for Conventional Relaying

---

1: $L_{0,0}^t = \text{load}(\text{UTs} \in \text{BS}, t)$
2: **for** $i = 1$ to 4 **do**
3:     $L_{1,i}^t = \text{load}(\text{FRN}_i, t)$
4:     $L_{2,i}^t = \text{load}(\text{UTs} \in \text{FRN}_i, t)$
5: **end for**
6: $C = (\sum_{i,j} L_{j,i}^{t-1})^{-1}$
7: **switch** (*phase*):
8:     **case** *first*:
9:         **for** $i = 1$ to 4 **do**
10:             $\text{slot}_t^{\text{BS} \rightarrow \text{FRN}_i} \sim CL_{1,i}^{t-1}$
11:         **end for**
12:     **case** *second*:
13:         $\text{slot}_t^{\text{FRN}_1 \rightarrow \text{UTs}} = \text{slot}_t^{\text{FRN}_2 \rightarrow \text{UTs}} \sim C(L_{2,1}^{t-1} + L_{2,2}^{t-1})$
14:         $\text{slot}_t^{\text{FRN}_3 \rightarrow \text{UTs}} = \text{slot}_t^{\text{FRN}_4 \rightarrow \text{UTs}} \sim C(L_{2,3}^{t-1} + L_{2,4}^{t-1})$
15:     **case** *third*:
16:         $\text{slot}_t^{\text{BS} \rightarrow \text{UTs}} \sim CL_{0,0}^{t-1}$
17: **end switch**

---

maximisation for conventional relaying (Algorithm 1) is based on such a scheduler, which exploits additional feedback regarding the buffer load $L_{j,i}^{t-1}$ in the previous cycle $t - 1$. It means that the total length of the multi-frame remains unchanged, however, the lengths of specific frames, corresponding to the duration of the service time, and referred to as slots, are calculated adaptively [22], [10]. The calculation is based on the buffer utilisation percentage $CL_{j,i}^{t-1}$ in the previous cycle, where C denotes the reciprocal of the total system load. Specifically, during the first phase, the length of slot$_t^{\text{BS} \rightarrow \text{FRN}_i}$, which the BS assigns to itself for the purposes of serving the FRN$_i$, $(i = 1 \ldots 4)$, is proportional to $CL_{1,i}^{t-1}$. The $L_{1,i}^t$ is defined as the buffer utilisation of the FRN$_i$ in the cycle t and is denoted as load(FRN$_i$, t). During the second phase, when the two spatially independent FRNs serve their UTs, the slot length for each such couple is calculated with respect to the average buffer utilisation of their respective UTs in the previous cycle. Finally, during the third phase, when the BS serves its UTs, the slot length becomes proportional to the buffer utilisation for these UTs in the previous cycle.

3.4 Manhattan Scenario Throughput

Following the description of adaptive approach to conventional relaying, the performance evaluation is provided in this section. For the purposes of simulation, each UT is assumed to generate traffic according to predefined parameters and there are random sessions established between distinct pairs of UTs. The lengths of time slots, according to the framing policy, are calculated by Algorithm 1, outlined in Section 3.3. Regarding the system parameters, there is an assumption made that during the simulation five UTs per Radio Access Point (RAP) may be active and each UT may set up a session during which it sends packets to another UT with steady intensity. In particular, the UTs of addresses 5 - 10, 11 - 15, 16 - 20 and 21 - 25 are assigned to the FRNs of addresses 1, 2, 3 and 4, respectively. In turn, the UTs of addresses 26 - 30 belong to the BS of address 0. The simulation termination condition is met when the packet of number 2,000,000 reaches its destination. The simulation results obtained for two different sizes of the buffer memory are presented below.

First, the simulation results for the system of buffer memory length equal to 30 packets are outlined for both non-adaptive (fixed slot length) and adaptive (dynamically adjusted slot length) case. In Figure 3.6, the total number of packets sent by

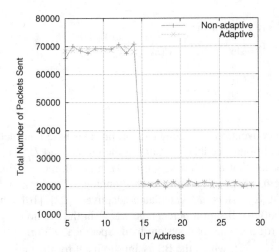

Fig. 3.6 Total number of packets sent for buffer length equal to 30

each UT is depicted. One should note that the packet generation intensity is on average the same for the non-adaptive and adaptive systems. In both cases, all the UTs belonging to FRN_1 and FRN_2 send much more packets than the UTs belonging to FRN_3 or FRN_4, or to the BS. Thanks to this disproportion, the behaviour of the adaptive approach can be better observed, as it is possible to assign bigger slot to the FRNs characterised by higher buffer memory utilisation. The results pertain-

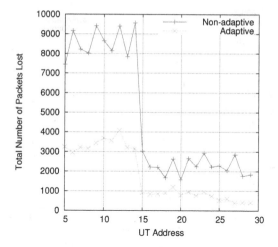

Fig. 3.7 Total number of packets lost for buffer length equal to 30

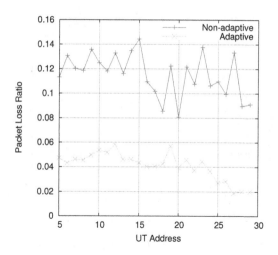

Fig. 3.8 Packet loss ratio for buffer length equal to 30

ing to this system are depicted in Figure 3.7 and Figure 3.8, where the number of packets lost and the packet loss ratio curves are presented. These results prove that adaptive conventional relaying is much more stable and the overall throughput is significantly maximised. A similar simulation investigation is performed for both the non-adaptive and adaptive systems with doubled length of the buffer memory, i.e. 60 packets. The packet generation intensity shown in Figure 3.9 is, on average, the same as in the previous case and also this time far better performance can be observed, as depicted in Figure 3.10 and Figure 3.11. However, the most important results are presented in Figure 3.12, where the non-adaptive system with buffer memory of length equal to 60 packets is compared to the adaptive one with

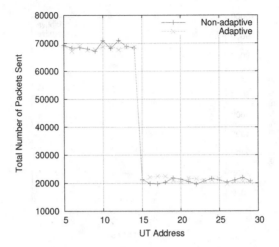

Fig. 3.9 Total number of packets sent for buffer length equal to 60

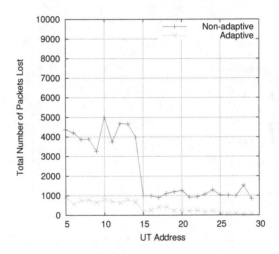

Fig. 3.10 Total number of packets lost for buffer length equal to 60

buffer memory length of 30 packets. The conclusion from this comparison is that the adaptive system, being equipped with twice less amount of the buffer memory, may perform not worse than the non-adaptive one.

Moreover, the average delay is computed between the time a packet is generated by a UT and the time this packet leaves the FRN. The results are presented in Table 3.3 and Table 3.4. The observed delay time reduction is even up to 60% in the case of the adaptive system with buffer memory of length of 60 packets. Yet another parameter evaluated during simulations is the difference between the average time slot length for both systems and both sizes of the buffer memory. The results are presented in Figure 3.13 and Figure 3.14. They show that the maximum average

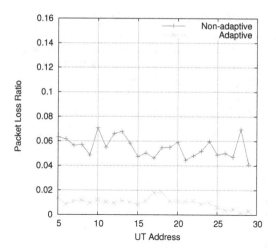

Fig. 3.11 Packet loss ratio for buffer length equal to 60

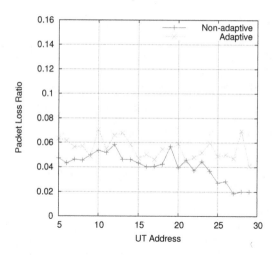

Fig. 3.12 Packet loss ratio comparison for buffer length equal to 30 and 60

Table 3.3 Average delay per FRN for buffer length equal 30

FRN	Average delay for the non-adaptive system [unit]	Average delay for the adaptive system [unit]	Gain [%]
1	98.8	57.3	42
2	103.4	60.7	41
3	93.6	55.4	41
4	95.9	57.1	41

Table 3.4 Average delay per FRN for buffer length equal 60

FRN	Average delay for the non-adaptive system [unit]	Average delay for the adaptive system [unit]	Gain [%]
1	163.5	67.8	59
2	176.1	70.4	60
3	150.0	71.5	52
4	153.4	70.5	54

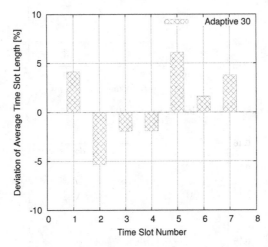

Fig. 3.13 Average time slot length deviation for adaptive system (buffer size 30)

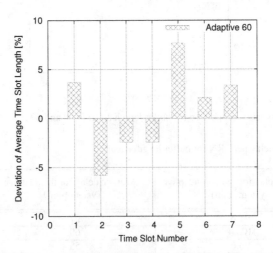

Fig. 3.14 Average time slot length deviation for adaptive system (buffer size 60)

variation in the time slot length is less than 8 per cent, which is rather not much in the context of the achieved throughput improvement and delay time reduction.

As presented in this section, the performance of the L3DF conventional relaying in a fixed environment may improved mostly at the network layer. Yet another dimension appears when the relay node may be selected in an ad hoc manner. This is especially visible for the cooperative relaying schemes to be described in the following sections, where additionally diversity becomes exploitable.

3.5 Virtual Antenna Arrays

The concept of Virtual Antenna Arrays (VAAs) was introduced by Dohler in [4], [5]. In this book, it is analysed together with Distributed Space-Time Block Coding (DSTBC), which may be perceived as an attempt to make use of the advantages of the idea of legacy space-time block coding (see Section 2.4) for the purposes of mapping them onto cooperative relaying. Under the name of distributed space-time block coding, this approach was first introduced by Laneman and Wornell in [16], where both the repetition-based cooperative diversity and space-time coded cooperative diversity techniques were analysed and compared. In particular, the system

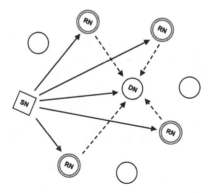

Fig. 3.15 Distributed space-time block coding

model shown in Figure 3.15 was analysed, where the process of transmission between the source node and the destination node again comprises two phases. During the first phase, the SN broadcasts its signal, which is received by the DN, as well as by the potential RNs. Afterwards, this signal is processed by these intermediate RNs and finally resent towards the DN during the second phase. This retransmission can be done either in the repetition-based or space-time coded manner. Although both these approaches are capable of achieving full spatial diversity in the number of cooperating terminals, the space-time coded diversity outperforms the repetition-based one, because it can be used more effectively for higher spectral efficiencies [16]. One should also note that there are different retransmission schemes known, when the second phase is considered [21]. For example, in [13], among others, the Simple Adaptive Decode and Forward (SAdDF) and Complex Adaptive Decode and Forward (CAdDF) protocols are briefly compared. The difference between the two pertains to the operation undertaken by each of them in case an RN is not able to relay the received signal. In the case of SAdDF [11], if a Relay Node decides not to participate in retransmission, it remains silent, whereas in the case of CAdDF [15], the source node retransmits the signal, instead. In [15] and [16], it is mentioned that the terminals share their antennas and other resources to create a virtual array through distributed transmission and signal processing. The idea of DSTBC seems then to naturally extend to the concept of Virtual Antenna Arrays. One should note, however, that VAAs do not have to employ space-time block coding and other spatio-temporal processing methods, such as space-time trellis coding (see Section 2.5), are equally well applicable.

In the case of Virtual Antenna Arrays, the transmission is also subdivided into phases. Similarly, during the first one, the SN delivers the signal to the DN, as well as to a set of the intermediate RNs. Then, during the second phase, the RNs resend the received signal towards the DN. Following, this concept was additionally extended to the distributed physical layer meshing [1], while the capacity of VAA-aided transmission was discussed in [2] (see also Section 2.2). Besides, the applicability of STTC was investigated in [6] (see also Section 2.5). Finally, the idea of Virtual Antenna Arrays was generalised to multi-hop systems in [3]. Such a Gen-

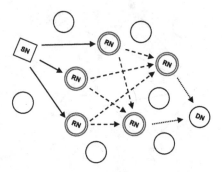

Fig. 3.16 Generalised virtual antenna array concept

eralised Virtual Antenna Array (GVAA) concept is presented in Figure 3.16 and it is based on the assumption that there may be a number of tiers of relay nodes between the source and destination ones. Consequently, there is no direct connection between the SN and DN, while the distinct tiers of RNs form separate VAAs. Now, one could even safely say that in the case of GVAA each VAA performs the operation of a DSTBC. Last but not least, either in the case of GVAA or any other approach to cooperative relaying, sufficiently tight synchronisation is definitely required. However, the issue of synchronisation remains beyond the scope of this work and perfect synchronisation will be assumed in the remainder of this book.

3.6 Equivalent Distributed Space-Time Block Encoder

Following the discussion of different approaches to space-time coded cooperative transmission, a simulation analysis is included into this section [22]. The aim of this analysis is to extend the considerations presented in Section 2.4 by evaluating the performance of Distributed Space-Time Block Coding in a fading channel to prepare the background for further considerations. To this end, first an Equivalent Distributed Space-Time Block Encoder (EDSTBE) is defined below [22].

Definition 3.1. A perfectly synchronised set of distributed relay nodes, connected to the source node via error-free links, and able to cooperatively encode the received signals according to a given space-time block code matrix X_Y, conceptually forms and is defined as an Equivalent Distributed Space-Time Block Encoder E_Y^X, where $X_Y = \{G_2, G_3, G_4, H_3, H_4\}^1$. In this context, the relay nodes are referred to as the transmitters, whereas the destination one, acting as a single receiving antenna, is referred to as the receiver.

When analysed in the context of cooperative relaying, the above assumption of error-free first-hop links, may seem a bit unrealistic. However, as it will be shown in Chapter 4, such an approximation has a lot of sense, when a real world scenario is investigated. Particularly, such a model proves reasonable when the transmission is initiated and coordinated by a Base Station or an Access Point. The reason is that such Radio Access Points, powered from electrical grid, can transmit using much higher power than battery powered mobile UTs, and therefore provide a much higher SNR.

Pursuant to Definition 3.1, three different Equivalent Distributed Space-Time Block Encoders: E_2^G, E_3^G, and E_4^G are evaluated and the number of the receivers is limited to 1. To this end, an MISO flat fading Rayleigh channel, characterised by lack of correlation among any of the wireless links between one of the transmitters and the receiver is used, where the fading coefficients are calculated according to a sum-of-sinusoids statistical simulation model given in [24]. This model is advantageous, because the autocorrelation and cross-correlation functions of the quadrature components, as well as the autocorrelation function of the complex envelope, match the desired ones even if the number of sinusoids is a one digit number. According to this publication, the normalised fading process of a statistical sum-of-sinusoids model is given by (3.2):

$$X(t) = X_c(t) + jX_s(t) \tag{3.2}$$

where $X_c(t)$ is defined as (3.3):

$$X_c(t) = \frac{2}{\sqrt{K}} \sum_{k=1}^{K} \cos(\varphi_k) \cos(\omega_d t \cos \alpha_k + \phi) \tag{3.3}$$

---

[1] Other space-time block codes for complex signal constellations are, however, not excluded.

and $X_s(t)$ is defined as (3.4):

$$X_s(t) = \frac{2}{\sqrt{K}} \sum_{k=1}^{K} \sin(\varphi_k) \cos(\omega_d t \cos \alpha_k + \phi) \tag{3.4}$$

Here α_k is equal to (3.5):

$$\alpha_n = \frac{2\pi k - \pi + \theta}{4K}, \qquad k = 1, 2, \ldots, K \tag{3.5}$$

and random variables θ, ϕ and φ_k are statistically independent and uniformly distributed over the range of $[-\pi, \pi)$ for all k. As stated in [24], this model can be directly used for the purposes of generating multiple uncorrelated fading waveforms for frequency selective fading channels, Multiple Input Multiple Output (MIMO) channels, as well as diversity combining scenarios. This proves its direct applicability to the investigations carried out in this book.

In particular, the three following systems: $p = \{2, 1, 1, 1\}$, $p = \{3, 1, 1, 1\}$ and $p = \{4, 1, 1, 1\}$ are analysed, defined with the aid of the notation outlined in Section 3.2 [12]. Unlike in Section 2.4, where decision metrics of the general form were used (2.18), all the simulations reported here [22] are performed on the basis of modified metrics, provided in [19]. Particularly, when G_2 space-time block code is concerned, the metric (2.18) can be re-written as (3.6):

$$\sum_{j=1}^{M} \left(\left| r_1^j - h_{1,j} s_1 - h_{2,j} s_2 \right|^2 + \left| r_2^j + h_{1,j} s_2^* - h_{2,j} s_1^* \right|^2 \right) \tag{3.6}$$

The maximum-likelihood detection aims to find the minimum value of this metric for all the possible combinations of s_1 and s_2. Unfortunately, the calculation is computationally inefficient, when carried out according to this formula. It was shown, however, that a more efficient approach is feasible [19]. In particular, the expression (3.6) can be expanded so that the components independent of the code words may be disregarded. Consequently, the problem of minimisation becomes equivalent to minimising the following expression (3.7):

$$|s_1|^2 \sum_{j=1}^{M} \sum_{i=1}^{2} |h_{i,j}|^2 - \sum_{j=1}^{M} \left[r_1^j h_{1,j}^* s_1^* + \left(r_1^j \right)^* h_{1,j} s_1 + r_2^j h_{2,j}^* s_1 + \left(r_2^j \right)^* h_{2,j} s_1^* \right] +$$
$$+ |s_2|^2 \sum_{j=1}^{M} \sum_{i=1}^{2} |h_{i,j}|^2 - \sum_{j=1}^{M} \left[r_1^j h_{2,j}^* s_2^* + \left(r_1^j \right)^* h_{2,j} s_2 - r_2^j h_{1,j}^* s_2 - \left(r_2^j \right)^* h_{1,j} s_2^* \right] \tag{3.7}$$

This formula, in turn, may be perceived as composed of two parts, the first of them being exclusively the function of s_1 and the second one being exclusively the function of s_2. As a result, one may conclude that the total value of this metric is minimum, when each of its aforementioned two components is minimum. Finally, the metric for E_2^G code can be written as (3.8) [19]:

$$\left|\sum_{j=1}^{m} R_j - s\right|^2 + \left(-1 + \sum_{j=1}^{m}\sum_{i=1}^{2}|h_{i,j}|^2\right)|s|^2 \tag{3.8}$$

where for $s = s_1$ the signal R_j, received by antenna j, is given by (3.9):

$$R_j = r_1^j h_{1,j}^* + (r_2^j)^* h_{2,j} \tag{3.9}$$

and for $s = s_2$ the signal R_j, received by antenna j, is given by (3.10):

$$R_j = r_1^j h_{2,j}^* - (r_2^j)^* h_{1,j} \tag{3.10}$$

Similarly, the metric for E_3^G code can be then written as (3.11) [19]:

$$\left|\sum_{j=1}^{m} R_j - s\right|^2 + \left(-1 + 2\sum_{j=1}^{m}\sum_{i=1}^{3}|h_{i,j}|^2\right)|s|^2 \tag{3.11}$$

where for $s = s_1$ the signal R_j, received by antenna j, is given by (3.12):

$$R_j = r_1^j h_{1,j}^* + r_2^j h_{2,j}^* + r_3^j h_{3,j}^* + (r_5^j)^* h_{1,j} + (r_6^j)^* h_{2,j} + (r_7^j)^* h_{3,j} \tag{3.12}$$

for $s = s_2$ the signal R_j, received by antenna j, is given by (3.13):

$$R_j = r_1^j h_{2,j}^* - r_2^j h_{1,j}^* + r_4^j h_{3,j}^* + (r_5^j)^* h_{2,j} - (r_6^j)^* h_{1,j} + (r_8^j)^* h_{3,j} \tag{3.13}$$

for $s = s_3$ the signal R_j, received by antenna j, is given by (3.14):

$$R_j = r_1^j h_{3,j}^* - r_3^j h_{1,j}^* - r_4^j h_{2,j}^* + (r_5^j)^* h_{3,j} - (r_7^j)^* h_{1,j} - (r_8^j)^* h_{2,j} \tag{3.14}$$

and for $s = s_4$ the signal R_j, received by antenna j, is given by (3.15):

$$R_j = -r_2^j h_{3,j}^* + r_3^j h_{2,j}^* - r_4^j h_{1,j}^* - (r_6^j)^* h_{3,j} + (r_7^j)^* h_{2,j} - (r_8^j)^* h_{1,j} \tag{3.15}$$

Finally, the metric for G_4 code can be expressed as (3.16) [19]:

$$\left|\sum_{j=1}^{m} R_j - s\right|^2 + \left(-1 + 2\sum_{j=1}^{m}\sum_{i=1}^{4}|h_{i,j}|^2\right)|s|^2 \tag{3.16}$$

where for $s = s_1$ the signal R_j, received by antenna j, is given by (3.17):

$$R_j = r_1^j h_{1,j}^* + r_2^j h_{2,j}^* + r_3^j h_{3,j}^* + r_4^j h_{4,j}^* + (r_5^j)^* h_{1,j} + (r_6^j)^* h_{2,j} + (r_7^j)^* h_{3,j} + (r_8^j)^* h_{4,j}$$
$$(3.17)$$

for $s = s_2$ the signal R_j, received by antenna j, is given by (3.18):

$$R_j = r_1^j h_{2,j}^* - r_2^j h_{1,j}^* - r_3^j h_{4,j}^* + r_4^j h_{3,j}^* + (r_5^j)^* h_{2,j} - (r_6^j)^* h_{1,j} - (r_7^j)^* h_{4,j} + (r_8^j)^* h_{3,j}$$
$$(3.18)$$

for $s = s_3$ the signal R_j, received by antenna j, is given by (3.19):

$$R_j = r_1^j h_{3,j}^* + r_2^j h_{4,j}^* - r_3^j h_{1,j}^* - r_4^j h_{2,j}^* + (r_5^j)^* h_{3,j} + (r_6^j)^* h_{4,j} - (r_7^j)^* h_{1,j} - (r_8^j)^* h_{2,j}$$
$$(3.19)$$

and for $s = s_4$ the signal R_j, received by antenna j, is given by (3.20)[2]:

$$R_j = r_1^j h_{4,j}^* - r_2^j h_{3,j}^* + r_3^j h_{2,j}^* - r_4^j h_{1,j}^* + (r_5^j)^* h_{4,j} - (r_6^j)^* h_{3,j} + (r_7^j)^* h_{2,j} - (r_8^j)^* h_{1,j}$$
$$(3.20)$$

For evaluation purposes, it is guaranteed that the power transmitted by each transmit antenna (RN) is normalised so that the total transmitted power is equal to unity. Additionally, the signal is perturbed by the additive white Gaussian noise with zero mean and $N_0/2$ variance per dimension. Always 10 million bits are transmitted and the QPSK modulation scheme is used. The obtained results are depicted in Figure 3.17, where one can observe that the specific curves are characterised by different slope. This illustrates the aforementioned diversity gain, as discussed in Section 2.3.

3.7 Conclusion

In this chapter, the conventional and cooperative transmission protocols were described. After the analysis of the adaptive conventional relaying scenario of Manhattan type, the cooperative relaying based on Virtual Antenna Arrays was introduced together with Distributed Space-Time Block Coding as the preferred transmission method. This was supplemented by the definition and evaluation of the Equivalent Distributed Space-Time Block Encoder to be further referred to in applicable scenarios. In particular, the described concepts will be immediately complemented and extended in the following chapter, where routing information will be exploited for the purposes of the organization of cooperative transmission.

---

[2] When implementing this metric, the author of this book noticed that the formula given in [19] is erroneous and so is the one given in [20] on page 107. This error does not exist in [14], however, one should note that all the G_4 metrics therein also contain a typo and are written, as if they were pertaining to a code matrix composed of 3 columns (see index i on pages 87-88), whereas this code was designed for 4 transmitting antennas.

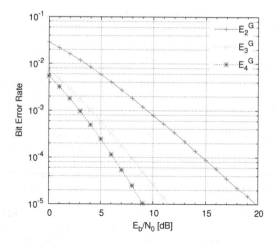

Fig. 3.17 E_2^G, E_3^G, and E_4^G performance comparison

References

1. M. Dohler and H. Aghvami. Distributed PHY-Layer Mesh Networks. *14th IEEE Personal, Indoor and Mobile Radio Communications, PIMRC*, pages 2543–2547, Sep. 2003.
2. M. Dohler, J. Dominguez, and H. Aghvami. Link capacity analysis for virtual antenna arrays. *IEEE 56th VTC 2002-Fall, Vehicular Technology Conference*, pages 440–443, Sep. 2002.
3. M. Dohler, A. Gkelias, and H. Aghvami. A resource allocation strategy for distributed MIMO multi-hop communication systems. *IEEE Communications Letters*, 8(2):99–101, Feb. 2004.
4. M. Dohler, E. Lefranc, and H. Aghvami. Virtual Antenna Arrays for Future Wireless Mobile Communication Systems. *International Conference on Telecommunications, ICT*, Jun. 2002.
5. M. Dohler and Y. Li. *Cooperative Communications - Hardware, Channel & PHY*. Wiley, 2010.
6. M. Dohler, B. Rassool, and H. Aghvami. Performance evaluation of STTCs for virtual antenna arrays. *The 57th IEEE Semiannual Vehicular Technology Conference, VTC*, pages 57–60, Spring 2003.
7. K. Doppler, S. Redana, M. Wódczak, P. Rost, and R. Wichman. Dynamic resource assignment and cooperative relaying in cellular networks: Concept and performance assessment. *EURASIP Journal on Wireless Communications and Networking*, Jul. 2007.
8. N. Esseling, R. Pabst, and B. Walke. Delay and Throughput Analysis of a Fixed Relay Concept for Next Generation Wireless Systems. *The 11th European Wireless Conference*, pages 273–279, Apr. 2005.
9. N. Esseling, B. Walke, and R. Pabst. Performance Evaluation of a Fixed Relay Concept for Next Generation Wireless Systems. *15th IEEE International Symposium on Personal, Indoor and Mobile Radio Communications, PIMRC*, Sep. 2004.
10. M. Głąbowski and M. Wódczak. On throughput maximization oriented approach to buffer memory management in context of the relay-based Manhattan-type deployment concept. *IST Mobile Summit*, Jun. 2006.
11. P. Herhold, E. Zimmermann, and G. Fettweis. A Simple Cooperative Extension to Wireless Relaying. *2004 International Zurich Seminar on Communications, Zurich, Switzerland*, Feb. 2004.

12. P. Herhold, E. Zimmermann, and G. Fettweis. On the Performance of Cooperative Amplify-and-Forward Relay Networks. *ITG Conference on Source and Channel Coding (SCC), Erlangen, Germany*, Jan. 2004.
13. P. Herhold, E. Zimmermann, and G. Fettweis. Cooperative multi-hop transmission in wireless networks. *Computer Networks Journal*, 49(3):299–324, Oct. 2005.
14. M. Jankiraman. *Space-Time Codes and MIMO Systems*. Artech House, 2004.
15. J. N. Laneman, D.N.C Tse, and G.W. Wornell. Cooperative diversity in wireless networks: Efficient protocols and outage behavior. *IEEE Transactions on Information Theory*, 50(12):3062–3080, Dec. 2004.
16. J. N. Laneman and G. W. Wornell. Distributed space-time-coded protocols for exploiting cooperative diversity in wireless networks. *IEEE Transactions on Information Theory*, 49(10):2415–2425, Oct. 2003.
17. A. Raha, N. Malcolm, and W. Zhao. Hard Real Time Communications with Weighted Round Robin Service in ATM Local Area Networks. *1st International Conference on Engineering of Complex Computer Systems*, pages 96 – 104, 1996.
18. D. C. Schultz, B. Walke, R. Pabst, and T. Irnich. Fixed and Planned Relay Based Radio Network Deployment Concepts. *10th Wireless World Research Forum*, Oct. 2003.
19. V. Tarokh, H. Jafarkhani, and A. R. Calderbank. Space-time block coding for wireless communications: performance results. *IEEE Journal on Selected Areas in Communications*, 17(3):451–460, Mar. 1999.
20. B. Vucetic and J. Yuan. *Space-Time Coding*. John Wiley & Sons, 2003.
21. M. Wódczak. On the Adaptive Approach to Antenna Selection and Space-Time Coding in Context of the Relay Based Mobile Ad-hoc Networks. *XI National Symposium of Radio Science URSI, Poznań, Poland*, pages 138–142, Apr. 2005.
22. M. Wódczak. *On Routing information Enhanced Algorithm for space-time coded Cooperative Transmission in wireless mobile networks*. PhD thesis, Faculty of Electrical Engineering, Institute of Electronics and Telecommunications, Poznań University of Technology, Poland, Sep. 2006.
23. M. Wódczak. Autonomic Cooperative Networking for Wireless Green sensor Systems. *International Journal of Sensor Networks (IJSNet)*, 10(1/2), 2011.
24. Y. R. Zheng and C. Xiao. Simulation Models with Correct Statistical Properties for Rayleigh Fading Channels. *IEEE Transactions on Communications*, 51(6):920–928, Jun. 2003.
25. E. Zimmermann, P. Herhold, and G. Fettweis. On the Performance of Cooperative Diversity Protocols in Practical Wireless Systems. *58th VTC Orlando*, Fall 2003.
26. E. Zimmermann, P. Herhold, and G. Fettweis. On the Performance of Cooperative Relaying in Wireless Networks. *European Transactions on Telecommunications*, 16(1):5–16, Jan.-Feb. 2005.

Chapter 4
Routing Information Enhanced Cooperative Transmission

4.1 Introduction

The idea of performing cooperative relaying with the use of Virtual Antenna Arrays and on the basis of Distributed Space-Time Block Coding seems very appealing and beneficial, as discussed in the previous chapter. The question arises, however, how to enable and organise such cooperation among devices in a networked system. This issue is addressed with the aid of the proactive Optimised Link State Routing Protocol featuring the Multi-Point Relay selection heuristic. In particular, it is shown how Virtual Antenna Arrays may be seamlessly integrated into such an existing protocol and then the necessary modifications are outlined together with certain algorithmic extensions, as well as performance and overhead analysis is carried out. The presented solution already displays readiness for being integrated into the bigger picture of an autonomic cooperative system design framework, what will be exploited in the next chapter.

4.2 Optimised Link State Routing Protocol

The Optimised Link State Routing protocol [5], [9] was primarily designed for Mobile Ad-hoc Networks (MANETs) [14]. Such environments are usually characterised by very dynamic changes in network topology, and, therefore, the protocol should be tailored accordingly so that, keeping the overhead at a reasonable level, it would be able to follow these changes and provide accurate routing information. Generally, there are three fundamental routing concepts [1], [12] known for MANETs. First of all, there is a proactive approach where each network node performs topology recognition on a regular basis, so the routing tables are always up-to-date. Unfortunately, unless optimised, this approach may be costly in terms of protocol overhead. Secondly, one may distinguish the reactive approach, where topology recognition is performed once the routing table needs to be updated. Hence, the protocol overhead is reduced, but, in turn, the delay related to route selection, increases. Last but not least is the hybrid approach combining the advantages of the aforementioned methods, depending on the activity of mobile nodes in specific regions of the network. As long as the topology changes are rather insignificant, the reactive attitude may be more appropriate, otherwise the proactive one is used.

In this book, special emphasis is laid on the Optimised Link State Routing (OLSR) protocol. Not only does OLSR belong to the proactive class, but it also features the Multi-Point Relay (MPR) selection heuristic. This heuristic aims to reduce the protocol overhead understood as the number of control messages broadcast for the purposes of network topology information dissemination [5], [11]. Generally, the idea is to transmit the Topology Control (TC) messages exclusively through the selected neighbour nodes, which belong to the one-hop neighbourhood of a given node and have been selected to cover the whole strict two-hop neighbourhood of this node. Such one-hop neighbours are recognised with Hello messages, which are received by each of them, but are not retransmitted. Hello messages, generated on the basis of the information stored in the Local Link Set, Neighbour Set and MPR Set [5], are broadcast by nodes on all their interfaces in a periodic manner. The operation of link sensing is necessary for the purposes of detecting whether a radio link exists in both directions, merely in one or even none of them. There is a direct association between the existence of a link and the existence of a neighbour. Therefore, Hello messages allow each node to discover both its entire one-hop and two-hop neighbourhoods, while the data gathered with their aid are exploited by the MPR selection heuristic.

In order to provide sufficient context for the OLSR protocol extensions to be presented later in this chapter (see Section 4.6), below the formats of the OLSR packet and Hello message are briefly described on the basis of [5]. In fact, also the above-mentioned Topology Control messages are encapsulated in OLSR packets but their description is not provided here and, instead, the reader is referred directly to [5]. As depicted in Figure 4.1, each OLSR packet starts with the Packet Length field (16

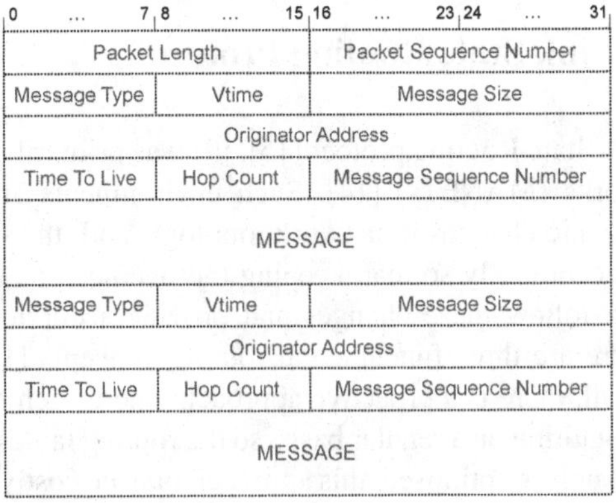

0 ... 7	8 ... 15	16 ... 23	24 ... 31
Packet Length		Packet Sequence Number	
Message Type	Vtime	Message Size	
Originator Address			
Time To Live	Hop Count	Message Sequence Number	
MESSAGE			
Message Type	Vtime	Message Size	
Originator Address			
Time To Live	Hop Count	Message Sequence Number	
MESSAGE			

Fig. 4.1 OLSR packet format

bits) specifying its length in bytes. It is followed by the Packet Sequence Number field (16 bits), which is incremented by one each time a new OLSR packet is transmitted. Then distinct messages follow, preceded by a header containing a number of fields. First, there is the Message Type (8 bits) indicating the type of the carried message. One should note that the size of this field is sufficient to make future attempts at defining new message types possible. Second is the Vtime field (8 bits), also known as Validity time, which defines for how long the received information is to be considered valid in case there is no update to it in the meantime. This time is represented in the form of the mantissa a (four most significant bits) and the exponent b (four least significant bits), and based on this, the target validity time can be calculated according to the following formula (4.1) (also compare formula 4.3):

$$V_t = C\left(1 + \frac{a}{16}\right)2^b \tag{4.1}$$

where C is a constant scaling factor assumed to be equal to (4.2) [5]:

$$C = \frac{1}{16} = 0.0625s \tag{4.2}$$

Next is the Message Size field (16 bits) containing the size of the message in bytes, as counted from the beginning of a given Message Size field until the beginning of the next Message Size field, or in case there are no more messages, until the end of the OLSR packet. What follows is the Originator Address field (32 bits) with the main address of the node being the original issuer of this message. It is also crucial to note [5] that this address does not correspond to the Source address of the Internet Protocol (IP) header, which is changed each time to the address of the intermediate interface retransmitting this message. Then, there is the Time To Live (TTL) field (8 bits) pointing out the maximum number of hops a given message may be retransmitted. It is decremented by 1 before retransmission occurs and a given message may not be retransmitted, if its TTL is equal to 0 or 1. Following comes the Hop Count field (8 bits) indicating the number of hops the packet has traversed so far, as well as the Message Sequence Number field (16 bits) containing a unique identification number, exploited for the purposes of ensuring that a specific message is transmitted once only. Finally, the MESSAGE field of variable size contains the relevant message, such the Hello one.

As depicted in Figure 4.2, Hello messages also comprise a number of important fields [5]. Firstly, there is the Reserved field (16 bits) which must be set to 0000000000000000[1]. It is followed by the Htime field (8 bits), also known as the Holding time, which is used for specifying the Hello message emission interval over a given interface. This interval is represented in the form of the mantissa a (four most significant bits) and the exponent b (four least significant bits). Based on this, the emission interval can be calculated according to the following formula (4.3) (see also formula 4.1):

---

[1] In the case of the specification [5], 13 zeros are given instead of 16, while there is no point in leaving 3 of them unset.

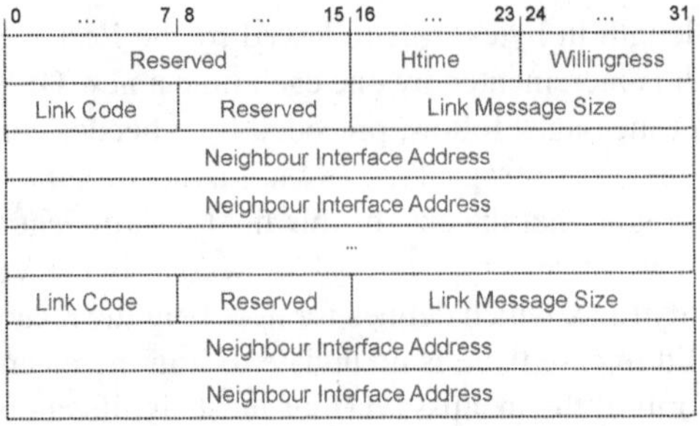

0 ... 7	8 ... 15	16 ... 23	24 ... 31
Reserved		Htime	Willingness
Link Code	Reserved	Link Message Size	
Neighbour Interface Address			
Neighbour Interface Address			
...			
Link Code	Reserved	Link Message Size	
Neighbour Interface Address			
Neighbour Interface Address			

Fig. 4.2 Hello message format

$$H_t = C\left(1 + \frac{a}{16}\right) 2^b \qquad (4.3)$$

where C is a constant scaling factor assumed to be equal to (4.2). Although the predefined Hello message emission interval amounts to 2 seconds, it can range from 62.6 milliseconds up to almost 2.28 hours. Next, there is a Willingness field (8 bits) specifying whether a given node is willing to carry and forward traffic to other nodes or not. There are the following levels of willingness available: WILL_NEVER (0), WILL_LOW (1), WILL_DEFAULT (3), WILL_HIGH (6) and WILL_ALWAYS (7). One should note that, as long as Willingness is set to 0, a given node must never be selected as MPR. On the contrary, in case the willingness is equal to 7, such a node must always be selected as an MPR. Afterwards, comes the Link Code (8 bits) defining the type of the link between an interface of a given node and the listed interfaces of its neighbours, as well as the neighbour type, as depicted in Figure 4.3. Currently 16 different combinations are recognised, however, future extensions are

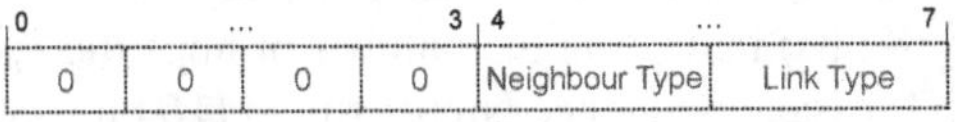

0	...		3	4	...		7
0	0	0	0	Neighbour Type		Link Type	

Fig. 4.3 *Link Code* format

also possible and, in fact, this option will be exploited later in this chapter. This field is structured so each the Neighbour Type and the Link Type field is assigned two bits. Following, all the specified Link Type and Neighbour Type values are given in Table 4.1 and Table 4.2, respectively. One should also note that a symmetric link is defined as a verified bi-directional link between two OLSR interfaces, whereas an asymmetric link is defined as link between two OLSR interfaces but verified in one direction only [5]. Last but not least appears the Neighbour Interface Address (16 bits) denoting the address of an interface of a given neighbour node.

Table 4.1 Link types

Link Type	Value	Description
UNSPEC_LINK	0	Indicates that no information about given links is specified.
ASYM_LINK	1	Indicates that given links are asymmetric which means that they are only heard.
SYM_LINK	2	Indicates that given links are symmetric.
LOST_LINK	3	Indicates that given links have been lost.

Table 4.2 Neighbour types

Neighbour Type	Value	Description
NOT_NEIGH	0	Indicates that given nodes are no longer considered as or have not yet become symmetric neighbours of this node.
SYM_NEIGH	1	Indicates that there exists at least one symmetric link between this node and each of the listed neighbours.
MPR_NEIGH	2	Indicates that there exists at least one symmetric link between this node and each of the listed neighbours, additionally selected as MPRs.

4.3 Multi-Point Relay Station Selection Heuristic

One of the main advantages of the Optimised Link State Routing protocol is its ability to use the selected nodes only for the purposes of the control data dissemination. These nodes are called Multi-Point Relays (MPRs), and they are chosen by a given node x out of its all symmetric one-hop neighbours. In consequence, all other neighbours in the range of this node, which do not belong to its MPR Set, also receive and process the control messages this node broadcasts, but do not retransmit them (Figure 4.4). Such an approach aims to minimise the number of redundant retransmissions and so to optimise the global control traffic. In order to perform the MPR selection heuristic, the node x must first collect all the necessary information regarding its one-hop and two-hop neighbourhoods. To this end, it exploits the data acquired through the reception of the aforementioned Hello messages, periodically transmitted by its one-hop neighbours. More specifically, each one-hop neighbour n of this node advertises its one-hop neighbourhood, as well as the status of the corresponding links. Consequently, the node x can identify both its symmetric neighbourhoods and then perform the MPR selection heuristic. In fact, there are three different neighbourhood types [5], as outlined in Table 4.3.

Prior to outlining the MPR selection heuristic [5], let us define $N(x)$ as the set of one-hop neighbours and $N^{(2)}(x)$ as the set two-hop neighbours of a given node x. Let us also define $MPR(x)$ as the set of Multi-Point Relays of this node x, where a Multi-Point Relay is a node which was selected by its one-hop neighbour x to

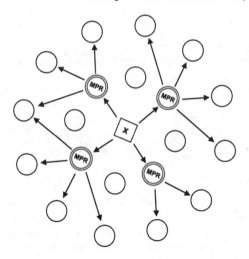

Fig. 4.4 Multi-point relaying

Table 4.3 Neighbourhood type

Neighbourhood Type	Definition
Symmetric one-hop neighbourhood of the node x	A set of nodes which have at least one symmetric link to the node x.
Symmetric two-hop neighbourhood of the node x	A set of nodes, excluding the node x itself, which have a symmetric link to the symmetric one-hop neighbourhood of the node x.
Symmetric strict two-hop neighbourhood of the node x	A set of nodes, excluding the node x and its neighbours, which have a symmetric link to a symmetric one-hop neighbour of the node x, characterised by the willingness different from WILL_NEVER.

retransmit all the broadcast messages that it receives from this node, provided that a message to be retransmitted is not a duplicate and its Time To Live (TTL) field carries value greater that one [5]. The MPR selection heuristic is performed with the use of both the sets of one-hop and two-hop neighbours. First, node x includes in the $MPR(x)$ set these of its symmetric one-hop neighbours n that are the only ones to provide reachability to a node n^2, located in the strict symmetric two-hop neighbourhood, and additionally are always willing to carry and forward traffic [11]. Next, while there still exist any uncovered nodes in $N^2(x)$ the heuristic keeps on selecting this node n in $N(x)$, which has not been inserted into the $MPR(x)$ set so far, and is characterised by the highest willingness to carry and forward traffic. In the case of multiple choices, the one is chosen which provides the highest reachability $R(n)$, i.e. through which the highest number of still uncovered nodes in $N^2(x)$ may be reached. Otherwise, if it is impossible to select one node only, the node with the highest degree is chosen, where the degree $D(n)$ of a one-hop neighbour denotes the number of its symmetric neighbours, excluding all the members of $N(x)$ and the node x performing the computation [5]. Once the MPR selection procedure is completed, Topology Control messages can be disseminated solely via this limited set of identified MPR nodes and as a result the protocol overhead may be significantly reduced [11].

4.4 Integration of Virtual Antenna Arrays

The OLSR protocol is well suited to large and dense mobile networks. This feature, together with its proactive flavour and link state nature, makes OLSR an ideal candidate for incorporating the concept of Virtual Antenna Arrays [6]. In fact, it resulted in the development of the Routing information Enhanced Algorithm for Cooperative Transmission (REACT) [15], [16]. REACT is based on the classic MPR selection heuristic, and it facilitates the process of the organisation of VAA-aided cooperative transmission. The main idea is to execute the classic MPR selection heuristic iteratively to identify nodes which can act together as VAAs. Additionally, extra MPR sets are created, ready to be exploited, if adaptive increase in protocol overhead is necessary to guarantee its proper functioning [5]. Of course, one needs to remember that, typically, relay nodes may cooperate at the Link layer according to more or less sophisticated schemes (see Chapter 3). However, their knowledge about the network topology and the parameters of separate radio links is limited to the closest, one-hop neighbourhood only. While it is still possible to imagine a more complex approach, able to collect additional details at the Link layer, it seems way more straightforward to utilise the information readily available at the Network layer instead.

In fact, such a goal may be achieved with the aid of the OLSR protocol which, thanks to its inherent mechanisms, allows each of the nodes to acquire the knowledge about their one-hop and two-hop neighbourhoods. What is more, it is possible to identify these one-hop neighbours in $N(x)$ which can provide connectivity to some two-hop neighbours in $N^{(2)}(x)$. This is one of the main reasons for basing REACT on the MPR selection heuristic, as there exists an obvious common aspect between the two. Namely, only these nodes are identified as MPRs which provide connectivity to a two-hop neighbour $n^{(2)}$. This assumption also holds true for the nodes to be pre-selected as the VAA set elements. An example is given in Figure 4.5, where it is shown that the nodes identified as Multi-Point Relays can also function as Mobile Relay Nodes and therefore form a Virtual Antenna Array.

4.5 Algorithmic Description

Following the notation introduced in Section 4.3, based on additional link-state information provided by the OLSR protocol, the REACT algorithm attempts to assign

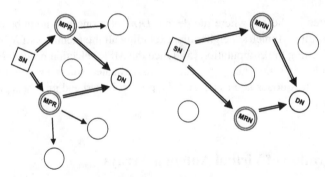

Fig. 4.5 Common aspect of the MPR selection heuristic and the VAA technology

RNs to specific VAAs [15], [19]. Let us just recall that the sets $N(x)$ and $N^{(2)}(x)$ are formed by one-hop neighbours and two-hop neighbours of node x, respectively. It is also assumed that both these sets contain symmetric nodes only reachable via bi-directional links. Moreover, the degree of a symmetric one-hop neighbour is defined as the number of its symmetric neighbours, excluding all the members of $N(x)$ and node x itself [5]. Analysing Algorithm 2, first, each neighbour n characterised

---

Algorithm 2 REACT

1: **for all** $n \in N(x)$ **do**
2:     **if** $degree(n) = 0$ **then**
3:         $N(x) \leftarrow N(x) \backslash \{n\}$
4:     **end if**
5: **end for**
6: $i \leftarrow 1$
7: **while** $N(x) \neq \emptyset$ **do**
8:     $MPR^i(x) \leftarrow$ OLSR_MPR_HEURISTIC($N(x)$)
9:     **for all** $n \in MPR^i(x)$ **do**
10:         **for all** $n^{(2)} \in N^{(2)}(x)$ **do**
11:             **if** $n = neighbour(n^{(2)})$ **then**
12:                 $VAA(x, n^{(2)}) \leftarrow VAA(x, n^{(2)}) \cup \{n\}$
13:             **end if**
14:         **end for**
15:     **end for**
16:     $N(x) \leftarrow N(x) \backslash MPR^i(x)$
17:     $i \leftarrow i + 1$
18: **end while**

---

by zero degree ($degree(n) = 0$) is removed by node x from set $N(x)$. Then the classic MPR selection heuristic is executed iteratively over set $N(x)$, until all the potential MPR nodes have been assigned to distinct $MPR^i(x)$ sets. Consequently, each iteration results in an additional MPR set, i.e. secondary, ternary and so on, and then all nodes it contains, are allocated to the most relevant Virtual Antenna Arrays. These VAAs are denoted as $VAA(x, n^{(2)})$ and are capable of providing cooperative connectivity between the source node x and the destination node $n^{(2)}$, where $n^{(2)}$ belongs to the two-hop symmetric neighbourhood of node x. As a result, any intermediate node n can be included in more than one VAA. In this way not only all the intermediate nodes are pre-selected, but also additional redundancy is introduced to the MPR selection mechanism, as it was signalled previously. Such redundancy can be utilised in case there appear very sudden changes in the network topology so these additional $MPR^i(x)$ can be taken into account autonomically to provide better coverage [17], [18].

In the case of the Optimised Link State Routing protocol, all the one-hop neighbours are notified about having been chosen as MPRs with the aid of Hello messages. The same pattern is followed in the case of informing them about having been assigned to the specific VAAs. In this way additional information is conveyed upon which a node n can learn, firstly, that it is supposed to take part in space-time coded cooperative transmission and, secondly, according to which column of the relevant space-time block code it should process the received signal. Evaluation of

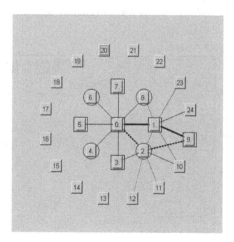

Fig. 4.6 REACT scenario

the performance of the proposed approach is carried out in the scenario depicted in Figure 4.6 [15], [19]. The wireless network is formed by the nodes of the following types: SN (0), RN (1-8) and DN (9-24). Relay nodes in squares are the ones that would be selected by the classic Multi-Point Relay selection heuristic, whereas the ones in circles belong to the redundant, secondary MPR set, selected additionally by the REACT algorithm. For the purposes of reducing the complexity of the simulated system, the maximum size of VAA is limited to 2, so the unity rate, G_2 space-time block code, is applicable [3], [13]. Consequently, once the REACT mode is active, the following primary $MPR^1(0) = \{1,3,5,7\}$ and secondary $MPR^{(2)}(0) = \{2,4,6,8\}$ sets are created, respectively. Actually, this is where the readiness of the proposed solution for the incorporation into autonomic system design [21], [4] is clearly visible, as the latter set may be used autonomically, if an increase in the control overhead is required. Based on both these sets and on the initial assumption that the test data stream would be originated from the SN and destined to the DN number 9, the $VAA(0,9) = \{1,2\}$ is set up at the beginning of the simulation, as shown in Figure 4.6. However, since the RNs and the DN are assumed mobile and moving at the speed of 5 km/h, the assignment of RNs to $VAA(0,9)$ needs to be dynamic during the course of the simulation time. Although the simulator supports switching between the conventional two-hop and the REACT mode, it is guaranteed that at least two RNs are available in the region of interest, so that the space-time coded cooperative transmission is continuous. The simulation investigations are carried out on the downlink and the block SIMO (at the first hop) and MISO (at the second hop) Rayleigh channels are used [15], [19]. The channel coefficients for the links between distinct pairs of nodes are generated according to formulas proposed in [22], as already introduced in Section 3.6. Similarly to all previous simulations, the total transmitted power, either by a single node or a VAA, is always normalised to unity. Additionally, the signal is perturbed by the additive white Gaussian noise with zero mean and $N_0/2$ variance per dimension. Always 100 million bits are transmitted and the QPSK modulation scheme is used. Under these

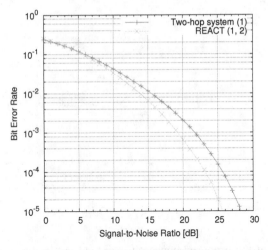

Fig. 4.7 Comparison of the performance for both the REACT and the conventional two-hop system

simulation assumptions, both the conventional two-hop mode without cooperative relaying, where the transmission is assisted by one RN only, and the REACT mode, exploiting two RNs, are analysed. The corresponding results are presented in Figure 4.7, where the numbers placed in the legend next to the names of the specific system configurations denote the next hop neighbour(s).

Looking at the presented results, one might notice that the Bit Error Rate (BER) curves almost overlap in the region of low SNR values. This is undoubtedly related to the problem of the propagation of the first hop errors during the second hop, as analysed in [15]. Namely, the first hop transmission, where the SN feeds the selected

RNs over the SISO radio links, is less robust to the radio channel impairments when compared to the space-time coded cooperative transmission at the second hop (see also Section 2.4 and Section 3.6). Usually, however, the SN is represented by a Base Station or an Access Point. It means that higher transmission power and better antenna gains are available compared to battery powered Mobile Relay Nodes (see additionally Section 4.7). To quantify the influence of the first-hop transmissions,

---

Algorithm 3 Pre-selection of first hop neighbours

1: **for all** $n \in N(x)$ **do**
2:     **while** $j \geq 0$ **and** $P_x^{VAA(x,n^{(2)})[j]} < P_x^n$ **do**
3:         $VAA(x,n^{(2)})[j+1] \leftarrow VAA(x,n^{(2)})[j]$
4:         $j \leftarrow j - 1$
5:     **end while**
6: **end for**
7: $VAA(x,n^{(2)})[j+1] \leftarrow n$

Algorithm 3 is validated. This algorithm goes through the entire set $N(x)$ and promotes these potential RNs which are characterised by the highest received power P_x^n. In particular, two cases are evaluated, where the first-hop SNR value is maintained either at the level of 10 or 20 dB. The obtained results are depicted in Figure 4.8

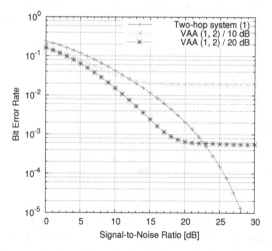

Fig. 4.8 Performance for first-hop SNR maintained at the level of 10 and 20 dB

where the numbers placed in the legend next to the names of the specific system configurations denote the next hop neighbour(s). This analysis aims to answer the question of what BER one could expect in the case of an equivalent dynamic system, if the SN would be able to pre-select RNs observing the received power level P_0^n being respectively 10 and 20 dB higher than the mean noise power N. Consequently, the performance of the investigated system would be not worse than what is indicated by the curves in Figure 4.8. Indeed, a gain in BER is observed, becoming the higher the higher the guaranteed SNR is. What is important is that such a system starts saturating at the values close to the assumed 10 and 20 dB, which additionally shows that the first hop is critical here.

4.6 Modifications to OLSR Protocol

The proposed concept requires certain extensions and modifications to the Optimised Link State Routing protocol [15], [19]. Special attention has been paid to make any changes compliant with the OLSR specification [5]. This assumption is fully met in the case of the first of them, related to the introduction of a new Neighbour Type, where an unallocated Neighbour Type value is exploited. This modification is required for the purposes of VAAs configuration, and more specifically, once the VAA pre-selection phase has been completed, each of the chosen RNs must be informed about its assignment to the specific $VAA(x, n^{(2)})$ set. In this way the RNs can conclude the way they are supposed to process the signals received from the SN. To this end, the list of Neighbour Types, originally specified by the OLSR protocol and given in Table 4.2, is extended by the introduction of a new VAA_NEIGH type, as described in Table 4.4. Consequently, Hello messages are able to convey Link messages of a new class, determined by this Neighbour Type. This information will have to be stored in an additional information repository, however, the details will be given after the Modified Hello message format has been discussed first. The introduction of a Modified Hello message format is inevitable, especially

Table 4.4 New neighbour type

Neighbour Type	Value	Description
VAA_NEIGH	3	Indicates that there exists at least one symmetric link between this node and each of the listed neighbours, additionally selected as VAAs.

for the purposes of extending the proposed REACT algorithm as outlined in Section 4.7. The Extended REACT [16] makes use of a more detailed information regarding the parameters of the specific radio links. Unfortunately, since the OLSR protocol was developed for Mobile Ad-hoc Networks [2], it collects merely very rough information about the links to the discovered neighbours (see Section 4.2). As a result, the parameters it provides would not be accurate enough to make the aforementioned extension feasible. Specifically, only four different Link Types are available, i.e. UNSPEC_LINK, ASYM_LINK, SYM_LINK and LOST_LINK, as previously summarised in Table 4.1.

Such an approach would sound reasonable, taking into account solely the characteristics of MANETs, where the knowledge whether a link is symmetric or not, suffices for setting up the communications. However, further optimisation of REACT demands a more detailed information pertaining to the power level of the received signal. In the case of the classic OLSR protocol, the Link Type is specified in the Link Code field of Hello message (see Figure 4.2). Consequently, each node groups in one Link Message these Neighbour Interface Addresses which are characterised by the same Neighbour Type and the same Link Type (see also Figure 4.3). Since information about link types is not very precise, such an approach guarantees that a number of Neighbour Interface Addresses are likely to fall into the same Link Message and it is why this kind of grouping seems rather effective, at least when the size of Hello messages is concerned [19]. Introducing any additional data regarding the link parameters might, on the one hand, spoil this effectiveness, because the more precise information is included, the smaller gets the number of nodes to be assigned to the same group. Eventually, each Link Message may contain one Neighbour Interface Addresses only. On the other hand, if the initial assumption regarding backward compatibility with the OLSR protocol specification is to be fulfilled, this might be the only reasonable solution to this issue. Therefore, a modified Hello message format is proposed in this section, which is depicted in Figure 4.9. This modified format contains a new 16 bit Extended Link Code field, compris-

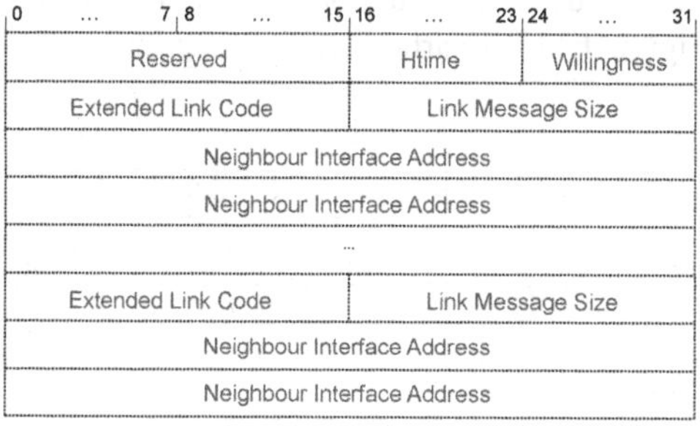

0 ... 7	8 ... 15	16 ... 23	24 ... 31
Reserved		Htime	Willingness
Extended Link Code		Link Message Size	
Neighbour Interface Address			
Neighbour Interface Address			
...			
Extended Link Code		Link Message Size	
Neighbour Interface Address			
Neighbour Interface Address			

Fig. 4.9 Modified Hello message format

ing both the classic Link Code and Reserved fields. The structure of this Extended Link Code is outlined in Figure 4.10. The idea is to utilise the four Most Signifi-

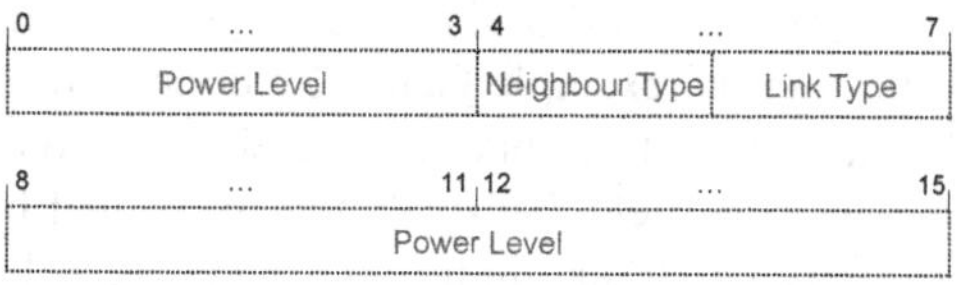

0 ... 3	4 ...	7
Power Level	Neighbour Type	Link Type

8 ... 11	12 ... 15
Power Level	

Fig. 4.10 Extended Link Code format

cant Bits (MSBs) of the Link Code field together with the eight additional bits of the Reserved field. It makes twelve bits in total which form together a new Power Level field, and are now available to convey additional information about the power level of the radio signal. As a result, the node x will be able not only to find out, whether its one-hop neighbour n can hear the transmitted signal coming from this node, but it may also learn what is the precise power level of this signal. Moreover, as Hello message sent by a node n usually contains similar information pertaining also to other one-hop neighbours of this node, which in turn may by the two-hop neighbours of the aforementioned node x, this node x can have far more concrete overview of the link parameters in its entire one-hop and two-hop neighbourhoods, especially if radio channel reciprocity could be assumed.

However, this very modification to the Hello message format is not completely transparent to the internal mechanisms of the Optimised Link State Routing protocol. Namely, unlike it was the case for the first modification, where it was sufficient to make the protocol aware of the new VAA_NEIGH type, here situation seems somewhat more complicated. The main problem is that the Reserved field is utilised which, according to the specification, should remain unchanged [5]. Moreover, the aforementioned four MSBs are exploited which are meant for future extensions, however, are not straightforwardly applicable in the case of the proposed modification. Therefore, for the purposes of overcoming this problem, a simplified attempt to guarantee backward compatibility is outlined below. Namely, when a Hello message is processed for the purposes of performing the classic protocol operations, the new Extended Link Code field should be masked with the Extended Link Code Mask, depicted in Figure 4.11. In particular, any specific implementation of such a

Fig. 4.11 Extended Link Code Mask format

modified OLSR protocol would have to utilise the defined mask for the purposes of performing a logical AND operation over all the Extended Link Codes, included in a specific Modified Hello message. The only exception to this rule is when the Power Level data need to be accessed for the needs of the Extended REACT algorithm. Such an approach seems to be the safest way from the backward compatibility perspective. However, one could also consider the introduction of a new Hello message format [15], [19]. Such an example Generalised Hello message is presented in Figure 4.12. Its format differs from both the classic and modified ones because it lacks the Link Message Size field. This solution is dictated by the fact that, as mentioned earlier, it is very likely that, in most of the cases, solely one Neighbour Interface Address will be included in one Link Message. Therefore, the size of the

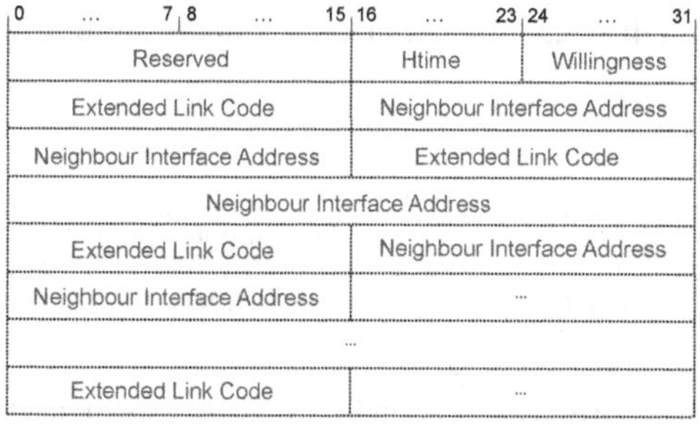

0 ... 7	8 ... 15	16 ... 23	24 ... 31
Reserved		Htime	Willingness
Extended Link Code		Neighbour Interface Address	
Neighbour Interface Address		Extended Link Code	
Neighbour Interface Address			
Extended Link Code		Neighbour Interface Address	
Neighbour Interface Address		...	
...			
Extended Link Code		...	

Fig. 4.12 Generalised Hello message format

entire Hello message could be reduced by skipping the Link Message Size field and always including only one Neighbour Interface Address in a Link Message[2].

Now, going back to the new VAA_NEIGH neighbour type, the initial idea, straightforwardly applicable to the original REACT, was to exploit the order in which the Neighbour Interface Addresses of VAA_NEIGH type are located on the list of the Neighbour Interface Addresses. This was meant for the purposes of notifying these neighbours about the way they are supposed to cooperate during the retransmission phase. More specifically, the position on the list would determine the identification number of the column, in the relevant space-time block code matrix, according to which each of the nodes belonging to $VAA(x, n^{(2)})$ should cooperatively process the retransmitted signal. The reciprocal of the number of addresses on such a list would originally specify the power scaling factor. One should note that in the Extended REACT the power will be not scaled in this sense anymore (Section 4.7). This way or another, since Modified Hello messages carry more detailed information about the parameters of distinct links, it is rather unlikely that two Neighbour Interface Addresses would fall into one Link Message. Therefore, it should be rather guaranteed that the order of the one-element Link messages would correspond to the identification numbers of the columns in the relevant space-time block code matrix. One should also take into account that, although unlikely, it might happen that two Neighbour Interface Addresses of VAA_NEIGH type fall into the same Link message after all. For this reason, once again, the optimum solution would be to include one Neighbour Interface Addresses of a VAA_NEIGH type in a Link Message only and, what is more, to place all such messages in Hello message in the first order to avoid fragmentation. This additionally proves, however, that the Generalised Hello message would be more applicable here.

Last but not least, after having processed such a Modified Hello message, each node VAA_NEIGH must store the acquired data [15], [19]. To this end, an additional VAA Selector Set is proposed to be maintained in the Neighbour Information Base[3]. This new VAA Selector Set is then formed by VAA-Selector Tuples of the

---

[2] Another issue is if such a structure could be still named a Link Message.

[3] For further information regarding the OLSR protocol repositories the reader is referred directly to the specification [5].

format presented in Table 4.5. As a result, each node can easily determine if it is

Table 4.5 VAA-Selector Tuple

Item	Description
VS_main_addr	The main address of a node which has selected this node as the element of a VAA.
VS_elem_id	The VAA element identification number specifying the column of the relevant space-time block code, according to which the retransmitted signals should be processed.
VS_time	The time at which this tuple expires and must be removed.

to cooperate after receiving a user data packet from a neighbour by simply comparing its address with VS_main_addr. If so, then it will use the relevant column of the space-time block code, as specified by VS_elem_id.

4.7 Algorithm Extension and Protocol Overhead

In this section, a low-mobility and high-density hot-spot scenario is investigated [15], [16]. In particular, Mobile Relay Nodes are taken into account but either the fixed or movable ones are of course conceptually not excluded. The velocity of mobile terminals is in-between 0 - 5 km/h which is equivalent to 0 - 1.4 m/s. A Line of Sight (LOS) channel model is assumed, characterised by the path-loss parameter $L(d)$, given by the formula (4.4) [8], [15]:

$$L(d) = 13.4\log_{10}(d) + 36.9 \quad \text{[dB]}, \tag{4.4}$$

where d represents the distance, in meters, and is limited in the following way: 5 m $< d <$ 29 m. The shadow fading standard deviation σ is equal to 1.3 dB. Additionally, the SN, being a Base Station in this case, is assumed to transmit with the average power equal to 200 mW. The BS is equipped with an antenna characterised by the gain of 8 dBi. The DNs, here UTs, which also act as mobile relays, are characterised by the corresponding parameters equal to 200 mW and 0 dBi. The downlink is investigated and the Quadrature Phase Shift Keying (QPSK) modulation scheme is assumed. Moreover, the noise figure introduced by the radio frequency chain of the mobile station is equal to 7 dB.

The analysed network topology, limited to one sector, is presented in Figure 4.13. The destination UT (18) is located 29 meters away from BS (0). The distance between BS and each UT belonging to the first (1 - 6) and to the second group (7 - 17) is equal to 10 and 24 meters, respectively. One should remember that BS and UTs

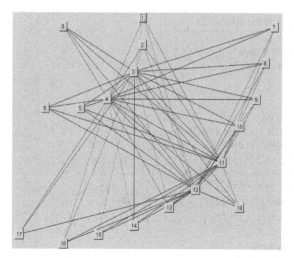

Fig. 4.13 Extended REACT scenario

are equipped with antennas offering different gains. Consequently, even if the power
level of the signal received from BS by the destination UT (18) is acceptable, it does
not necessarily hold true in the opposite direction. This is to some extent in contrast
with MANETs, for which the OLSR protocol was originally designed [20]. More
precisely, in the case of OLSR, if two nodes hear each other, a symmetric link can
be established without any additional considerations regarding the power levels, because in general, these nodes are perceived homogeneous [7]. For the needs of the
following analysis, it is assumed, however, that in the case of the considered scenario, a power threshold must be satisfied by the specific UT, if, using the OLSR
terminology, it is to be recognised as a neighbour of BS. As UT (18) does not meet
this requirement, it is assigned to the $N^{(2)}(x)$ set, whereas all the intermediate UTs
form the $N(x)$ set.

Similarly to Algorithm 2, also in the case of the Extended REACT, first each node
n having zero degree ($degree(n) = 0$) is removed by node x from the one-hop neighbour set $N(x)$, as outlined in Algorithm 4. Then the classic MPR selection heuristic, described in Section 4.3, is carried out iteratively over the set $N(x)$, until all the
nodes it contains have been assigned to the redundant $MPR^i(x)$ sets and pre-selected
into the most relevant Virtual Antenna Arrays $VAA(x, n^{(2)})$. After each iteration, all
the elements of set $MPR^i(x)$ are removed from set $N(x)$. However, the pre-selection
itself is performed in a different way. In the case of the original REACT, the order of potential relays in the specific $VAA(x, n^{(2)})$ sets was strongly correlated with
the MPR heuristic selection criteria, and therefore, not necessarily optimum. In the
proposed extension, additional information about the power levels of the signals
received by distinct nodes from their one-hop neighbours, collected by the modified OLSR protocol, is exploited. This information is stored in the Power Level
field of the Extended Link Code (see Section 4.6). Based on it, a node is placed
in $VAA(x, n^{(2)})$ at this position which corresponds to the power level at which it is

---

Algorithm 4 Extended REACT

1: **for all** $n \in N(x)$ **do**
2:    **if** $degree(n) = 0$ **then**
3:       $N(x) \leftarrow N(x) \backslash \{n\}$
4:    **end if**
5: **end for**
6: $i \leftarrow 1$
7: **while** $N(x) \neq \emptyset$ **do**
8:    $MPR^i(x) \leftarrow$ OLSR_MPR_HEURISTIC$(N(x))$
9:    **for all** $n \in MPR^i(x)$ **do**
10:       **for all** $n^{(2)} \in N^{(2)}(x)$ **do**
11:          **if** $n = neighbour(n^{(2)})$ **then**
12:             $j \leftarrow size(VAA(x, n^{(2)})) - 1$
13:             **while** $j \geq 0$ **and** $P^{n^{(2)}}_{VAA(x,n^{(2)})[j]} < P^{n^{(2)}}_n$ **do**
14:                $VAA(x, n^{(2)})[j+1] \leftarrow VAA(x, n^{(2)})[j]$
15:                $j \leftarrow j - 1$
16:             **end while**
17:             $VAA(x, n^{(2)})[j+1] \leftarrow n$
18:          **end if**
19:       **end for**
20:    **end for**
21:    $N(x) \leftarrow N(x) \backslash MPR^i(x)$
22:    $i \leftarrow i + 1$
23: **end while**

---

heard by the destination node. It means that the first relay nodes in a VAA set provide the best connectivity to the target UT. In this way the process of pre-selection is additionally optimised which is validated by simulations, as presented below.

A dedicated simulation environment is used [15] and the situation presented in Figure 4.13 is evaluated, where all the recognised neighbours of nodes 3, 4, 11 and 12 are depicted. The simulations are performed in the presence of a zero mean additive white Gaussian noise characterised by the power level N expressed in dBW. Always 100 million bits are transmitted. The noise power level is given in Figure 4.14 instead of SNR because the effective SNR in the system would vary from point to point, depending on the power level of the received signal modelled according to (4.4). The transmission is originated by BS (0) and destined to UT (18). Five distinct cases are considered. First, although it has been assumed that the destination UT is conceptually not a neighbour of BS, the performance of the one-hop link towards the destination UT is evaluated as the worst, reference case. Then two different configurations of a two-hop system are tested, where the transmission is assisted by UT (3) and UT (12), respectively. An advantage is observable, especially in the latter case, when the relaying UT is situated closer to the destination and the antenna gain of BS may be more efficiently exploited. Finally, REACT and its extended version are validated. In the case of the original REACT, where UT (3) and UT (4) are selected and consequently the $VAA(0, 18) = \{3, 4\}$ is configured, a significant improvement in the performance may be observed. What is even more important, the Extended REACT, featuring the modified pre-selection method, provides the ex-

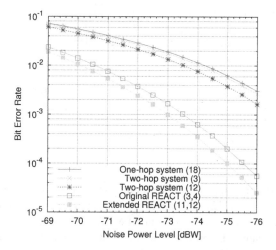

Fig. 4.14 Simulation results comparison for extended REACT

pected, additional performance gain. In its case, these MRNs (UTs) are pre-selected which are closer to the destination UT, and as a result the $VAA(0,18) = \{11,12\}$ is created. The detailed comparison of the results can be found in Figure 4.14, where the numbers placed in the legend next to the names of the specific system configurations denote the next hop neighbour(s).

Given the continually progressing convergence between the cellular systems and routing [20], especially in part related to the relay enhanced radio access network [7], it is also crucial to provide additional details about the process of routing when two-hop cooperative transmission is concerned. In general, when the Network layer sends a packet to a Destination Node denoted by an IP address of value R_dest_addr, it uses the IP address of value R_next_addr and requests the Link layer routines to send this packet, in a Medium Access Control (MAC) layer frame, to a MAC address corresponding to this R_next_addr IP address. Obviously, the MAC address is resolved with the aid of the Address Resolution Protocol (ARP) [10]. However, in the case of REACT a packet must be transmitted concurrently via all the RNs belonging to a given VAA. This issue may be addressed by associating an additional routing table with each column of the space-time block code matrix [15], which leads to a multidimensional routing table as depicted in Figure 4.15. For the sake of providing an example, let us follow the assumption that the size of the VAA is limited to two RNs, so that only two routing tables need to be maintained. Bearing in mind the Extended REACT configuration, RNs 11 and 12 are chosen to constitute the VAA serving the destination UT (18). According to the information stored in the VAA Selector Set, one entry will be included in the first routing table, whereas the other entry will go to the second routing table. In other words, when the Network layer routine at BS (0) is going to send a user data packet to UT (18) it should check both routing tables. As a result it would find out that cooperative transmission is possible, because two intermediate RNs are available: 11 and 12. It would then

1	R_dest_addr	R_next_addr	R_dist	R_iface_addr
1	R_dest_addr	R_next_addr	R_dist	R_iface_addr
1	R_dest_addr	R_next_addr	R_dist	R_iface_addr
2	R_dest_addr	R_next_addr	R_dist	R_iface_addr
3	R_dest_addr	R_next_addr	R_dist	R_iface_addr
...	...	...	...	...
k	R_dest_addr	R_next_addr	R_dist	R_iface_addr

Fig. 4.15 REACT routing table

request the underlying Link layer to send one packet to UT (18) via nodes 11 and 12, i.e. $VAA(0, 18) = \{11, 12\}$. The obvious requirement here is that this Link layer must be able to transmit frames to a group of MAC addresses.

As already indicated, the proposed solution integrates well with the routines of the Optimised Link State Routing protocol, so the instantiation of cooperative transmission does not require any additional messages to be transmitted. However, the format of the Hello Message is slightly modified in such a way that the normal operation of the OLSR should be not disturbed. This comes at a price in terms of an increase in the size of the Modified Hello Message, as mentioned in Section 4.6 [19]. It means that one may expect some overhead induced by the additional data that needs to be distributed (see Figure 4.16). The reason for that increase is mainly the

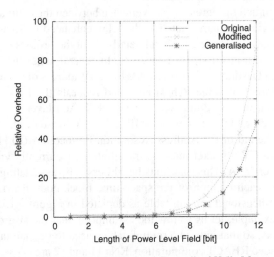

Fig. 4.16 Overhead introduced by the Modified and Generalised Hello Message format

Extended Link Code. Typically, the OLSR protocol distinguishes among 4 different link types and 3 different neighbour types. This gives 12 combinations which means that all the Neighbour Interface Addresses can be qualified to 12 groups, i.e. Link Messages, at most. For the proposed modification it suffices, when only some of these Link Messages are separated into smaller sets by including extra information about the power level. In particular, due to the type of the data required by the MPR selection heuristic and by the algorithms presented in this chapter, it is sufficient to focus on the SYM_LINK, as well as on the SYM_NEIGH and MPR_NEIGH. This limits the number of theoretically possible 16 combinations, in fact enlarged as a result of the introduction of the VAA_NEIGH, to just 2. The main factor influencing the overhead is then the size of the Power Level field [19]. This field is 12 bits long so there are theoretically 4096 values allowed which multiplied by the aforementioned 2 combinations gives 8192 possibilities. Taking into account that there are only singular interfaces characterised by a given power level, in the worst case, one would end up with 8192 Link Messages, each accompanied by a header of the length of 32 bits. This is of course the worst possibility and Figure 4.16 presents the expected overhead for different number of bits used. One can see that for 6 bits the overhead is almost diminishable.

4.8 Conclusion

In this chapter, the Routing information Enhanced Algorithm for Cooperative Transmission was presented as a solution for enabling Virtual Antenna Array aided cooperative relaying on the basis of Distributed Space-Time Block Coding and with the aid of the Optimised Link State Routing Protocol. Especially, the Multi-Point Relay station selection heuristic was employed and integrated with Virtual Antenna Arrays. To this end, certain modifications to the OLSR protocol were proposed keeping in mind backward compatibility. The introduced concept will be further integrated into the autonomic cooperative system design in the following chapter.

References

1. M. Abolhasan, T. Wysocki, and J. Lipman. Performance Investigation on three classes of MANET Routing Protocols. *Asia-Pacific Conference on Communications*, pages 774 – 778, Oct. 2005.
2. C. Adjih, E. Baccelli, and P. Jacquet. Link state routing in wireless ad-hoc networks. *IEEE Military Communications Conference, MILCOM*, pages 13–16, Oct. 2003.
3. S. Alamouti. A Simple Transmit Diversity Technique for Wireless Communications. *IEEE Journal on Selected Areas in Communications*, 16(8):1451–1458, Oct. 1998.
4. A.Liakopoulos, A.Zafeiropoulos, A.Polyrakis, M.Grammatikou, J.M.Gonzalez, M. Wódczak, and R.Chaparadza. Monitoring Issues for Autonomic Networks: The EFIPSANS Vision. *European Workshop on Mechanisms for the Future Internet*, 2008.
5. T. Clausen and P. Jacquet. Optimised Link State Routing Protocol (OLSR). *RFC 3626*, Oct. 2003.
6. M. Dohler and Y. Li. *Cooperative Communications - Hardware, Channel & PHY*. Wiley, 2010.

7. K. Doppler, S. Redana, M. Wódczak, P. Rost, and R. Wichman. Dynamic resource assignment and cooperative relaying in cellular networks: Concept and performance assessment. *EURASIP Journal on Wireless Communications and Networking*, Jul. 2007.

8. M. Dottling, W. Mohr, and A. Osseiran. *Radio Technologies and Concepts for IMT-Advanced*. Wiley, ISBN: 978-0-470-74763-6, December 2009.

9. P. Jacquet, P. Muhlethaler, T. Clausen, A. Laouiti, A. Qayyum, and L. Viennot. Optimised link state routing protocol for ad hoc networks. *IEEE International Multi Topic Conference*, pages 62–68, Dec. 2001.

10. D. C. Plummer. An Ethernet Address Resolution Protocol. *RFC 826*, Nov. 1982.

11. A. Qayyum, L. Viennot, and A. Laouiti. Multipoint Relaying for Flooding Broadcast Messages in Mobile Wireless Networks. *35th Annual Hawaii International Conference on System Sciences, HICSS*, Jan. 2002.

12. P. Sholander, A. Yankopolus, P. Coccoli, and S.S. Tabrizi. Experimental comparison of hybrid and proactive MANET routing protocols. *IEEE Military Communications Conference, MILCOM*, pages 513–518, Oct. 2002.

13. V. Tarokh, H. Jafarkhani, and A. R. Calderbank. Space-time block coding for wireless communications: performance results. *IEEE Journal on Selected Areas in Communications*, 17(3):451–460, Mar. 1999.

14. K. Weniger and M. Zitterbart. Mobile ad hoc networks - current approaches and future directions. *IEEE Network*, 8(4):52–59, Jul.-Aug. 2004.

15. M. Wódczak. *On Routing information Enhanced Algorithm for space-time coded Cooperative Transmission in wireless mobile networks*. PhD thesis, Faculty of Electrical Engineering, Institute of Electronics and Telecommunications, Poznań University of Technology, Poland, Sep. 2006.

16. M. Wódczak. Extended REACT - Routing information Enhanced Algorithm for Cooperative Transmission. *16th IST Mobile & Wireless Communications Summit 2007, Budapest, Hungary*, 1-5 July 2007.

17. M Wódczak. Aspects of Cross-Layer Design in Autonomic Cooperative Networking. *IEEE Third International Workshop on Cross Layer Design, Rennes, France*, 30 November - 1 December 2011.

18. M Wódczak. Autonomic Cooperation in Ad-hoc Environments. *5th International Workshop on Localised Algorithms and Protocols for Wireless Sensor Networks (LOCALGOS) in conjunction with IEEE International Conference on Distributed Computing in Sensor Systems (DCOSS), Barcelona, Spain*, 27-29 June 2011.

19. M. Wódczak. Autonomic Cooperative Networking for Wireless Green sensor Systems. *International Journal of Sensor Networks (IJSNet)*, 10(1/2), 2011.

20. M Wódczak. Convergence Aspects of Autonomic Cooperative Networking. *IEEE Fifth International Conference on Next Generation Mobile Applications, Services and Technologies, Cardiff, Wales, UK*, 14-16 September 2011.

21. M. Wódczak, T. B. Meriem, R. Chaparadza, K. Quinn, B. Lee, L. Ciavaglia, K. Tsagkaris, S. Szott, A. Zafeiropoulos, B. Radier, J. Kielthy, A. Liakopoulos, A. Kousaridas, and M. Duault. Standardising a Reference Model and Autonomic Network Architectures for the Self-managing Future Internet. *IEEE Network*, 25(6):50–56, November/December 2011.

22. Y. R. Zheng and C. Xiao. Simulation Models with Correct Statistical Properties for Rayleigh Fading Channels. *IEEE Transactions on Communications*, 51(6):920–928, Jun. 2003.

Chapter 5
Autonomic Cooperative System Design

5.1 Introduction

The concept of Routing information Enhanced Algorithm for Cooperative Transmission, proposed in the previous chapter and aiming to enable Virtual Antenna Array aided cooperative transmission with the use of the Multi-Point Relay station selection heuristic of the Optimised Link State Routing protocol, could be in general perceived as a more node-centric approach. The aim of this chapter is to put it in a wider context of an autonomic cooperative system design, where numerous cooperative and non-cooperative transmissions might be ongoing simultaneously. To this end, the Autonomic Cooperative Node is incorporated into the Generic Autonomic Network Architecture and certain managing entities are introduced such as the Cooperative Transmission Decision Element and Cooperative Re-Routing Decision Element. The proposed architectural extensions will form the basis for the analysis of autonomic cooperative network deployments in the next chapter.

5.2 Generic Autonomic Network Architecture

Autonomic networking has emerged as one of the most promising approaches toward the instantiation of the self-managing networked systems. This term seems very capacious, but it does not translate directly into the notion of being cognitive or autonomous. The primary role of autonomics is to instantiate network self-management without any explicit need for external intervention [5]. An autonomic system should then behave like a living organism, where a centralised entity monitors its status but refrains itself from interfering with it unless any such operation is definitely required. To map such a concept onto a networked system, one needs to apply specific network engineering mechanisms which are currently introduced onto the standardisation path [4], [23]. The idea of control loops is particularly applicable here because this way a Decision Element (DE) may control a Managed Entity (ME)

on the basis of a closed information flow and with the use of external monitoring
and policy related data. This is presented in Figure 5.1 describing the generic control loop. One should note that control loops and their respective DEs may in general

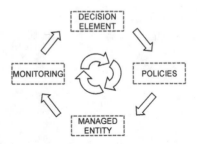

Fig. 5.1 Generic control loop

exist at the network, node, function, or protocol level (see also Figure 5.2). In this
way each of them has exclusive responsibilities, while they are still able to interact
among themselves. In other words, based on the control and monitoring information
available within a control loop, DEs are able to take their autonomic decisions which
might be to some extent affected by the data they exchange with their higher/lower
level, peering or sibling counterparts. Consequently, even a minor event may constitute a reason for triggering certain autonomic behaviour(s), potentially involving
a number of nodes or even the whole network. Autonomic networks are assumed
to be able to self-discover which may pertain to many aspects at the same time including, for example, service discovery, topology discovery, fault discovery, etc. In
order to avoid conflicting decisions negatively influencing the system performance,
there is a need for properly tailored hierarchical interactions among different DEs
to guarantee system stability and scalability. This is, in fact, very important for the
investigations related to autonomic cooperative behaviours [23].

5.3 Cooperation, Routing and Autonomics

To facilitate the integration of Autonomics into Routing Information Enhanced Cooperative Transmission (as defined in Chapter 4), specific architectural extensions,
well aligned with the rationale behind the Generic Autonomic Network Architecture (GANA) [5], are necessary. This is vital as the cooperative network might have
to be able to handle multiple concurrent cooperative and non-cooperative transmissions. The task of proper organisation of such transmissions is really demanding
because of the requirement for the knowledge of both the local and global scope parameters. Such parameters may translate into, on the one hand, the capabilities of a
mobile network to expose certain cooperative behaviours and, on the other hand, the
requirements with regard to expected QoS level. To accommodate such functional-
ity, it is primarily assumed that the definition of an Autonomic Node is enhanced

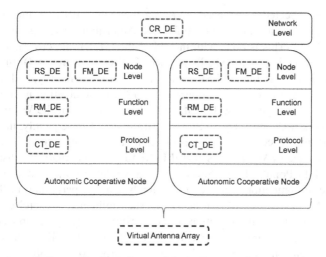

Fig. 5.2 Autonomic Cooperative Node from architectural perspective

with the notion of cooperative behaviours [20], [17]. Consequently, the Autonomic Cooperative Node is introduced to enable cooperation among the Managed Entities, orchestrated by their corresponding Decision Elements, as depicted in Figure 5.2. In particular, GANA defines the aforementioned four levels on which decision entities may appear. Starting from the protocol level, there is a new Cooperative Transmission Decision Element (CT_DE) introduced which is responsible for controlling the aspects of cooperative transmission protocol related to physical emulation of the Distributed Space-Time Block Coding [18]. It means that it is responsible for the action of processing the relayed signal according to the relevant column of the space-time block code matrix (see Section 5.5). The operation of CT_DE has to be aligned with the already existing Routing Management Decision Element RM_DE, located on the function level, which needs to interact with its sibling DEs, so the routing tables maintained at the cooperating nodes are properly synchronised [16]. What is more, the RM_DE also needs to act pursuant to the directions from the other existing DEs, i.e. the Resilience and Survivability Decision Element RS_DE and the Fault Management Decision Element FM_DE, both located on the node level [21]. In this case, the RS_DE is assumed to specifically cover aspects of service resilience and survivability. At the same time, it is supposed to interact with FM_DE which, in turn, controls the symptoms suggesting that a failure, for example in terms of service continuity, may be imminent. Finally, while all these DEs are located within Autonomic Cooperative Nodes, it is still necessary to provide substantial coordination on the network level. This task is accomplished by the Cooperative Re-Routing Decision Element CR_DE which is responsible for overseeing the situation from a higher level perspective and orchestrating transmissions so enhanced re-routing may be provided (see Section 5.6) [22].

As already indicated, there needs to be an interaction between the entities of GANA and the Routing information Enhanced Algorithm for Cooperative Transmission (REACT). This is done with the aid of the CT_DE directly managing the VAA enhanced version of the OLSR protocol [19] (see Section 4.4). In fact, the interaction with the Modified OLSR protocol is twofold. Primarily, and most naturally, it takes place through the adjustments to the dissemination interval of the Modified Hello messages, as well as through the adaptation of the Willingness of specific network nodes to carry and forward traffic [6]. Consequently, the convergence time of the process of VAA pre-selection can be tightly controlled and the nodes characterised by Willingness of WILL_NEVER are automatically disregarded as potential Multi-Point Relays. This fact is actually related to the second, additional aspect of the aforementioned interaction, namely the one pertaining to the influence on the MPR station selection heuristic. Such influence is exercised through the logic responsible for the selection of the transmission type, i.e. cooperative or non-cooperative. This logic is embedded into the Modified OLSR protocol through the REACT extension. As a result, the generic control loop (Figure 5.1) needs to be adequately adjusted, as depicted in Figure 5.3 [19]. In particular, the component

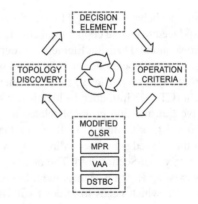

Fig. 5.3 Control loop for the investigated case

related to policies is narrowed down to operation criteria. The reason is that only does it seem more practical to talk about policies on the network level of the GANA hierarchy (see Figure 5.2), where the sense is generally more abstract. In the investigated case, the operation criteria pertain rather directly to instructing the OLSR protocol routines on how to act when cooperative behaviour is expected. Similarly, in the case of monitoring, the main activity translates into topology discovery and so it involves the collection of the data related to link types and their parameters. Topology information acquired this way may be directly applied in the process of VAA pre-selection and configuration. Consequently, the Modified OLSR protocol naturally becomes the Managed Entity driven by the Decision Element (Figure 5.3) [19]. The mainly affected components of the Modified OLSR protocol include its internal Multi-Point Relay station selection heuristic, enhanced with the external

Virtual Antenna Array entity using Distributed Space-Time Block Coding (i.e. the REACT module).

5.4 Neighbour Discovery and Address Auto-configuration

Proposing certain cooperative transmission related extensions to the Generic Auto-nomic Network Architecture (GANA), one needs to take into account a wider perspective of network self-configuration and self-discovery. In fact, GANA is fully integrated with IPv6 which provides certain relevant mechanisms, such as the Neighbour Discovery (ND) protocol. ND is exploited by network nodes, sharing the same link, not only to discover the presence of one another's, but also to learn the Link layer addresses of the other nodes, identify these willing to forward packets, maintain paths to the reachable neighbours, as well as detect the changed Link layer addresses [13]. Thanks to this mechanism, nodes should be able to discover themselves autonomically in new contexts. When a given node is no longer available, they may actively search for some active alternatives. In particular, after an interface has become active, a given node might start distributing router solicitation messages in order to stimulate other nodes to generate their router advertisements immediately. Based on the acquired information, the host can start address auto-configuration. More specifically, these nodes are routers which can use router advertisement messages to indicate whether a host should go for the stateless or stateful address auto-configuration.

In the case of the stateless approach, each host can generate its address as a combination of the sub-network and interface identifiers. This is very comfortable when one is not necessarily concerned about the specific addresses in use, of course as long as the ones assigned are unique and properly routable. In fact, there are a number of design goals given in [14], which define the scope of the stateless address auto-configuration. First of all, it is assumed that manual configuration should be not required and, in the basic approach, the address of an interface is formed on the basis of its Link layer address, combined with a prefix. Either small networks, consisting of a set of nodes sharing the same link, or large networks, consisting of a number of sub-networks, should not require stateful address configuration server. In the former case, all the addresses share the same prefix, whereas in the latter one each sub-network has its own prefix and routers advertise all the active prefixes.

The stateful solution is characterised by the existence of Dynamic Host Configuration Protocol (DHCP) servers, passing configuration parameters to network nodes. The configuration parameters are not limited to IPv6 addresses and may additionally include other information carried as options. In DHCPv6, clients and servers exchange messages with the use of User Datagram Protocol (UDP). DHCP clients transmit their messages to a reserved multicast address of a link scope, so they do not need to be configured with the address or addresses of DHCP servers [7]. In case the node and the DHCP server are not connected to the same link, a DHCP relay agent acts as an intermediary and takes care of proper delivery. The relay agent operates transparently, when viewed from the node's perspective. Once the node determines the address of the DHCP server, it might contact this server directly in some cases. What is important to note is that these mechanisms were designed for infrastructure-based environments and there are still a number of open issues regarding MANETs [3]. Specifically, the stateless address auto-configuration assumes that each node can communicate directly with all the other nodes, as if all the MANET nodes were connected to a single multicast link, which obviously needs not be the case [2]. The stateful approach, instead, assumes that each node is able to communicate either with the DHCP server or a DHCP relay, which again needs not be the case [2]. Additionally, one should take into account the lack of multi-hop support, as well as the lack of network merging and partitioning. A very detailed survey on this

Table 5.1 Address auto-configuration for standalone MANETs

Solution title	Brief characteristics
IP address Auto-configuration for Ad Hoc Networks	Random address selection from a range available for a given MANET, followed by detection of duplicate addresses.
IPv6 Auto-configuration in Large Scale Mobile Ad-Hoc Networks	Two DAD mechanisms are proposed: Strong DAD for the initial address auto-configuration phase and Weak DAD helping to avoid conflicts resulting from merging.
IP Address Assignment in a Mobile Ad Hoc Network	Dynamic address allocation based on the concept of binary split. Each node manages a disjoint address pool from which it can assign an address to another node without a need for consulting the neighbours.
An Address Assignment for the Automatic Configuration of Mobile Ad Hoc Networks	When a new node joins MANET then a given node may split the address pool it manages and assign the half of it to the node requesting an address.
No Overhead Auto-configuration OLSR	The Passive Duplicate Address Detection approach is used together with the OLSR protocol. Based on the fact that some protocol events occur generally for duplicate addressed and very rarely for the unique ones.
PDAD-OLSR: Passive Duplicate Address Detection for OLSR	Specific algorithmic approach is employed aimed at detecting duplicate addresses through the utilisation of different parameters contained in Hello and TC messages, as well as the addresses in OLSR protocol messages and IP headers.
Passive Duplicate Address Detection for On-demand Routing Protocols	Passive duplicate address detection is carried out on the basis of additional information contained in an on-demand protocol routing packets.
Prophet Address Allocation for Large Scale MANETs	A special function is used to determine the addresses of different nodes so the occurrence of the event of duplication is very unlikely. This way one can avoid non-unique address detection.
Address Auto-configuration in Optimised Link State Routing Protocol	Nodes periodically disseminate their addresses and a randomly generated sequence of bits of a fixed length. Duplicate addresses exist in case the identifiers do not match for a given address.

is contained in [3] and, following, Table 5.1 provides a brief outline of a selection of concepts from this work. In particular these solutions are presented which pertain to standalone MANET scenarios. For additional details and references to specific concepts, the reader is referred directly to [3]. Moreover, as this book targets mostly at the OLSR protocol related concepts, following three relevant approaches are described in a more detailed way.

First, the classic OLSR protocol is complemented by the No Overhead Auto-configuration extension (NOA-OLSR) [10]. In particular, the extension is used for initial address auto-configuration, and then for identifying conflicts with the addresses of other nodes, while the OLSR protocol operates in its usual manner. One should note that each node is allowed to fully run the OLSR protocol only after it has been confirmed that its address is unique. In the meantime, the node goes through different states and other nodes consider it as not fully reliable. The process of auto-configuration involves three stages: address generation, ongoing Duplicate Address Detection (DAD) and gradual entry into the OLSR network, as well as routing table contamination avoidance [10]. During the address generation stage, a given node monitors the exchanged messages and builds a list of the addresses in use. Then it selects a tentative address which should not be already on the aforementioned list, however, the procedure is not specified in detail. Following, the duplicate address detection process starts which consists in checking for inconsistencies in both the Hello and TC messages, as well as in the sequence numbers. Finally, the node gradually enters the OLSR network and more and more other nodes can use the messages it sends out. NOA-OLSR also checks if a given neighbour has been in the network long enough to successfully complete the DAD procedure and therefore be included in the routing table. The second solution, Passive Duplicate Address Detection (PDAD) for OLSR is based on a set of algorithms allowing the nodes to detect conflicts in the network through the observation of protocol anomalies [11], [1]. The general assumption is that some protocol events occur when duplicate addresses exist but are almost unlikely for unique ones. The nodes running PDAD obtain information about the state of the routing protocol daemon(s), running at other node(s), through the analysis of the incoming messages. Additionally, the node analysing an incoming message needs to be aware of the exact time the message was sent. Otherwise, some decisions could be outdated, as the state of the routing protocol daemon is continually changing over time. The aforementioned algorithms, which are used by network nodes for detecting duplicate addresses, make use of different parameters in Hello and TC messages, such as link states, link codes, message sequence numbers, as well as the addresses contained in OLSR protocol messages and in the IP headers. All the algorithms need to be implemented by a node so it is able to detect conflicts in the network. It is also assumed that duplicate addresses may affect the MPR selection heuristic [11]. Finally, the Address Auto-configuration in OLSR protocol is focused on guaranteeing address uniqueness for the case where different MANETs merge. The method assumes a distributed approach, where each node periodically sends out the list of all its addresses and the node identifier contained in a MAD (Multiple Address Declaration) message. In particular, first the address is assigned to an incoming node by means of random selection, control message exchange, etc. Then the DAD entity analyses the MAD messages and, in case a node finds its address on the list but the identifiers differ, a new address is chosen for this node. This problem usually occurs when networks merge. One should take into account that other nodes also detect such a conflict and once all of them started announcing it, there could appear a significant control overhead increase. For this reason it is better to let the MAD packets reach all the nodes in the network, rather than allow for the induction of the so called broadcast storm [9]. Additionally, the approach does not exclude optional elements such as route contamination avoidance or passive duplicate detection.

5.5 Cooperative Transmission Decision Element

This section aims to further build on top of the investigations provided in Chapter 4. In particular it is assumed here that in the cooperative mode, there might be a need to involve more nodes into one VAA. VAAs of increased size might be advantageous in terms of guaranteeing the required QoS[1]. As previously stated, it is REACT to interface internally with the MPR selection heuristic of the OLSR protocol. Now, however, OLSR as a whole is being directed by the protocol level CT_DE (see Figure 5.2) [17], [18]. In particular, CT_DE is responsible for informing the OLSR protocol about the necessity of involving additional intermediary nodes into VAAs. This is a consequence of the fact that the routines of the OLSR protocol have local scope only and it is the network layer CR_DE to request certain actions if a given part of a longer route, between distant end points, becomes a bottleneck [21]. From this angle, the CT_DE takes more a role of an interface here. Of course, to this end, the GANA decision entities need to continually gather the topology discovery information to ensure that Autonomic Cooperative Transmission (ACT) is at all possible. If not, then there might appear an option of Autonomic Cooperative Re-Routing (ACRR), as discussed in the following Section 5.6 [22].

Going back to the autonomic cooperative transmission case, the below analysis is performed with the aid of the Equivalent Distributed Space-Time Block Encoder and the notation introduced previously in Section 4.3 [16], [15]. Specifically, it is assumed that two-hop cooperative transmission takes place between the Source Node x and the Destination Node $n^{(2)}$. Additionally, the $C(n^{(2)})$ is defined as a set of channel coefficients for the radio channels between all n Relay Nodes, conceptually belonging to the one-hop neighbourhood $N(x)$ of node x, and the receiver $n^{(2)}$ located in its two-hop neighbourhood $N^{(2)}(x)$. Depending on the threshold value β, this transmission is assisted by at least three out of four intermediary Relay Nodes. These RNs form a Virtual Antenna Array which is denoted as $VAA(x, n^{(2)})$ and is able to encode the received signal in a distributed manner, i.e. with the use of selected space-time block codes as described in Section 2.4. Consequently, $VAA(x, n^{(2)})$ may

---

[1] Bigger size of a Virtual Antenna Array (VAA) means lower code rate so there is a trade-off between the robustness towards the impairments induced by the radio channel and the transmission speed as outlined in Section 2.4.

be perceived as a set or group of Autonomic Cooperative Nodes expressing cooperative behaviour through the instantiation of Distributed Space-Time Block Encoding. In particular, the concept of the EDSTBE is followed and switching between the two E_3^G and E_4^G modes is allowed. One should note that other codes are also applicable and the two have been chosen because they are characterized by the same code rate equal to $\frac{1}{2}$, which makes the results directly comparable. The strategy outlined in

---

Algorithm 5 Autonomic Cooperative Transmission

1: $n = min(C(n^{(2)}))$
2: **if** $C(n^{(2)})[n] < \beta$ **then**
3:     $VAA(x,n^{(2)}) \leftarrow VAA(x,n^{(2)})\backslash\{n\}$
4:     $mode \leftarrow E_3^G$
5: **else**
6:     $mode \leftarrow E_4^G$
7: **end if**

Algorithm 5 can be summarised in the following way [16]. The E_3^G mode is selected if the minimum value out of the moduli of the channel coefficients is lower than the given threshold β. As a result, one of the four allowable RNs is no longer included $VAA(x, n^{(2)})$ and the remaining three other transmit signals in accordance with the E_3^G code matrix. Otherwise, all four RNs are used and the EDSTBE operates in the E_4^G mode. In this way the worst radio link is autonomically discarded until its parameters potentially meet the criterion of the selection algorithm again. However, even if it is the case, the instructions coming from the hierarchy of Decision Elements might not allow it.

The simulations [16], [15] assume the use of the MISO flat fading Rayleigh channel, where the fading coefficients for each wireless link between the relay and destination node are calculated according to the modified and optimised formulas given in [24] (see also Section 3.6). When it comes to other parameters, the power emitted by each of the not excluded Relay Nodes is normalised so that the overall transmitted power is always equal to unity. The received signal is perturbed by additive white Gaussian noise characterised by zero mean and $N_0/2$ variance per dimension. Always 40 million bits are transmitted and the QPSK modulation scheme is used. Example results for $\beta = 0.8$ in comparison with the reference curve for the E_4^G encoder are presented in Figure 5.4, where the achieved gain is about 1 dB. Following, the detailed results pertaining to the Bit Error Rate (BER) improvement at a specific E_b/N_0 for a given threshold β are presented in Figure 5.5. The achieved improvement may be observed regardless the E_b/N_0 value and the optimum results are located more or less between $\beta = 0.5$ and $\beta = 0.8$. For higher β values, there is only minor improvement. Finally, the percentage of the E_3^G encoder usage is presented in Figure 5.6. It is about 2 per cent for $\beta = 0.1$ and about 95 per cent for $\beta = 1.2$. It means that in the former case, the autonomic cooperative system works almost all the time in the E_4^G mode, while in the latter one mostly the three best radio links are selected and the signal is coded in accordance with the E_3^G code matrix.

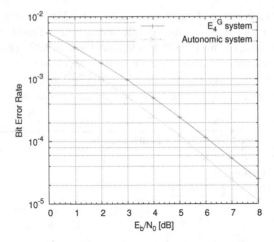

Fig. 5.4 BER comparison between the E_4^G encoder and the adaptive autonomic system for $\beta = 0.8$

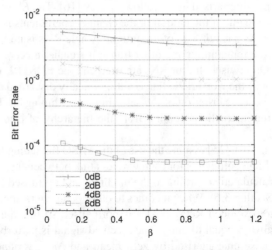

Fig. 5.5 Performance of the autonomic system in the function of β for the specific values of E_b/N_0

5.6 Cooperative Re-Routing Decision Element

As pointed out in Section 5.5, in some cases, the radio links leading from the SN towards the potential additional RNs and then also continuing to the DN might be of poor quality. If the relevant monitoring information, collected within a control loop during topology discovery, indicates this is the case, then Autonomic Cooperative Transmission (ACT) might not be sufficient. This information is passed by the Cooperative Transmission Decision Element (CT_DE) to the Cooperative Re-Routing Decision Element (CR_DE), so the latter may arrange for Autonomic Cooperative

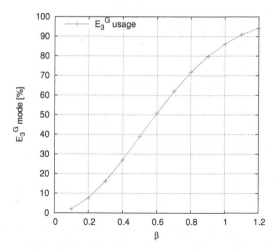

Fig. 5.6 Percentage of the E_3^G encoder usage

Re-Routing (ACRR) [22]. This process might be also supported by the Resilience and Survivability Decision Element (RS_DE), as well as by the Fault Management Decision Element (FM_DE) in certain cases [21].

The presented solution is an attempt to modify the existing approach to Fast Re-Route (FRR), so one could avoid the need for switching to alternative paths, as long as the transmission quality can be maintained. In other words, this operation may be postponed until the robustness of the transmission ongoing over the current path can be sufficiently increased with the aid of additional RNs [22]. To this end, the concept of Routing information Enhanced Algorithm for Cooperative Transmission (REACT) is exploited as outlined in Chapter 4, where the Multi-Point Relay (MPR) station selection heuristic [12] of the Optimised Link State Routing is applied for enabling VAA-aided cooperative transmission. Such a Modified OLSR protocol can be well integrated with FRR thanks to the possibility of routing over multiple paths. One should note, however, that the way multi-path routing is understood here, should not be confused with the Equal Cost Multipath Protocol (ECMP) [8]. While, in general, ECMP aims to help with load balancing, the ACRR improves robustness. Due to the nature of the wireless channel, it is characteristic for both MANETs and cooperative systems that the same data can be received from SN by a number of its neighbours, as a result of broadcasting [18]. Following, all of these neighbours which can reach the destination, simply resend the data. However, resending the data properly without any processing would not be possible, because the wireless channels towards the destination must be orthogonal, as explained in Chapter 2.

Figure 5.7 depicts the typical approach to FRR, where the packet stream between nodes A and C is routed over node B. In case there appears a failure of link 1 or link 2 or node B, the system needs to react properly. To this end, one of the readily available and pre-computed paths can be used instead almost immediately.

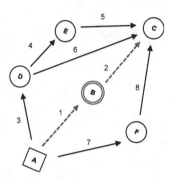

Fig. 5.7 Typical Fast Re-Route approach

In this very, case there are the following paths available, formed of the links (3, 4, 5) or (3, 6) or (7, 8). One should note that only one of them can be used. This pro-

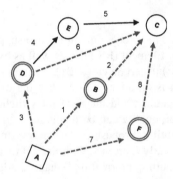

Fig. 5.8 Proposed approach

cess can be enhanced with cooperative transmission, as presented in Figure 5.8. As already mentioned, it is the nature of the wireless medium which helps the cooperative transmission to capitalise on the fact that data broadcast by node A is heard by nodes B, D and F. The reason for joining ACT and ACRR is that there is no need to wait with the exploitation of the alternative path(s) until a link failure has occurred, as described in Algorithm 6 [22]. The links originating from the Source Node x and

---

Algorithm 6 Autonomic Cooperative Re-Routing
1: **if** $((x,n) < \theta$ **or** $(n,n^2) < \theta)$ **then**
2:    **if** $VAA(x,n^{(2)}) \neq \emptyset)$ **then**
3:       route_cooperatively$(VAA(x,n^{(2)}))$
4:    **else**
5:       fast_reroute$((x,n^{(2)})$
6:    **end if**
7: **end if**

from the Relay Node n need to be checked and depending on whether both, or just one of them, may offer the requested transmission quality, further steps are undertaken. And so the algorithm needs to check whether there exist any group of nodes that would be able to form a VAA between the source and destination. The VAA sets are provided directly by REACT (see Section 4.5) and are used to facilitate the operation of ACRR whenever possible. It is then sufficient here to check if a given VAA set is valid. If so, then data may be routed cooperatively over multiple paths using the gains described below. Otherwise, it is necessary to start re-routing.

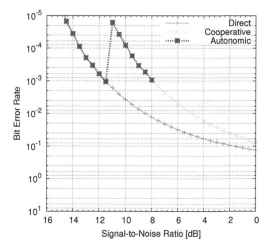

Fig. 5.9 Autonomic reaction at the threshold of BER = 0.001

To evaluate this approach, a series of simulations have been performed [22]. The corresponding results for BER = 0.001 are presented in Figure 5.9, for BER = 0.010 in Figure 5.10, and for BER = 0.100 in Figure 5.11. The obtained curves show that, for given Bit Error Rate (BER) thresholds, it is possible to autonomically switch from non-cooperative transmission (performed directly over one intermediary RN) to cooperative transmission with the use of a VAA, and this way observe a Signal to Noise Ratio (SNR) gain, potentially allowing to avoid the necessity of choosing another path. Specifically, it is possible to think about a situation where, based on the current MANET status, the CR_DE would normally force route change based on the options available from legacy FRR, but thanks to the advantages of ACRR it is still possible to keep the connection via a diversified set of paths despite some problems with single links. In other words, since alternative path(s) can be readily deployed, it is possible to additionally enable the aforementioned cooperative transmission. In particular, this concept is applicable in cases where there exist additional nodes that can form redundant paths of the same length (in terms of the number of hops) between the SN and DN.

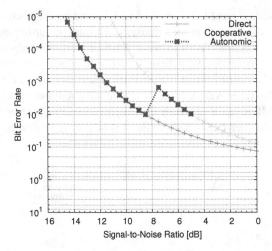

Fig. 5.10 Autonomic reaction at the threshold of BER = 0.010

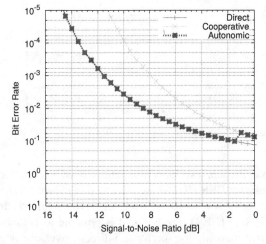

Fig. 5.11 Autonomic reaction at the threshold of BER = 0.100

5.7 Conclusion

In this chapter, the previously introduced concept of the Routing information Enhanced Algorithm for Cooperative Transmission was incorporated into the global framework of the Generic Autonomic Network Architecture. This required certain extensions to GANA itself, such as the inclusion of the notion of an Autonomic Cooperative Node, or the analysis of the new Cooperative Transmission Decision Element and Cooperative Re-Routing Decision Element. The outcome of these investigations will form the basis for the instantiation of cooperative autonomic network deployments to be introduced in the following chapter.

References

1. E. Baccelli. OLSR Passive Duplicate Address Detection. *draft-clausen-olsr-passive-dad-00*, July 2005.
2. E. Baccelli, K. Mase, S. Ruffino, and S. Singh. Address Autoconfiguration for MANET: Terminology and Problem Statement. *draft-ietf-autoconf-statement-04*, February 2008.
3. H. Moustafa C. Bernardos, M. Calderon. Survey of IP address autoconfiguration mechanisms for MANETs. *draft-bernardos-manet-autoconf-survey-05*, Jun. 2010.
4. R. Chaparadza, L. Ciavaglia, M. Wódczak, C.-C. Chen, B.A. Lee, A. Liakopoulos, A. Zafeiropoulos, E. Mancini, U. Mulligan, A. Davy, K. Quinn, B. Radier, N. Alonistioti, A. Kousaridas, P. Demestichas, K. Tsagkaris, M. Vigoureux, L. Vreck, M. Wilson, and L. Ladid. ETSI Industry Specification Group on Autonomic network engineering for self-managing Future Internet (ETSI ISG AFI). *10th International Conference on Web Information Systems Engineering, Poznań, Poland*, Sep. 2009. Published in Springer Lecture Notes in Computer Science (LNCS): Web Information Systems Engineering, Vol. 5802/2009, edited by G. Vossen, D. Long, and J. Yu.
5. R. Chaparadza, S. Papavassiliou, T. Kastrinogiannis, M. Vigoureux, E. Dotaro, K. A. Davy, M. Quinn, Wódczak, and A. Toth. Creating a viable Evolution Path towards Self-Managing Future Internet via a Standardizable Reference Model for Autonomic Network Engineering. *Chapter in the book: "Towards the Future Internet - A European Research Perspective" edited by G. Tselentis, J. Domingue, A. Galis, A. Gavras, D. Hausheer, S. Krco, V. Lotz, and T. Zahariadis, published by IOS Press, ISBN: 978-1-60750-007-0*, May 2009. Also published at the Future Internet Assembly 2009 in Prague.
6. T. Clausen and P. Jacquet. Optimised Link State Routing Protocol (OLSR). *RFC 3626*, Oct. 2003.
7. R. Droms, J. Bound, B. Volz, T. Lemon, C. Perkins, and M. Carney. Dynamic Host Configuration Protocol for IPv6 (DHCPv6). *RFC 3315*, Jul. 2003.
8. C. Hopps. Analysis of an Equal-Cost Multi-Path Algorithm. *RFC 2992*, Feb. 2004.
9. A. Laouiti. Address autoconfiguration in Optimized Link State Routing Protocol. *draft-laouiti- manet-olsr-address-autoconf-01*, July 2005.
10. K. Mase and C. Adjih. No Overhead Autoconfiguration OLSR. *draft-mase-manet-autoconf-noaolsr-01*, April 2006.
11. K. Mase and K. Weniger. PDAD-OLSR: Passive Duplicate Address Detection for OLSR. *draft-weniger-autoconf-pdad-olsr-01*, June 2006.
12. A. Qayyum, L. Viennot, and A. Laouiti. Multipoint Relaying for Flooding Broadcast Messages in Mobile Wireless Networks. *35th Annual Hawaii International Conference on System Sciences, HICSS*, Jan. 2002.
13. W. Simpson T. Narten, E. Nordmark. Neighbor Discovery for IP Version 6 (IPv6). *RFC 2461*, Dec. 1998.
14. S. Thomson and T. Narten. IPv6 Stateless Address Autoconfiguration. *RFC 2462*, Dec. 1998.
15. M. Wódczak. On the Adaptive Approach to Antenna Selection and Space-Time Coding in Context of the Relay Based Mobile Ad-hoc Networks. *XI National Symposium of Radio Science URSI, Poznań, Poland*, pages 138–142, Apr. 2005.
16. M. Wódczak. *On Routing information Enhanced Algorithm for space-time coded Cooperative Transmission in wireless mobile networks*. PhD thesis, Faculty of Electrical Engineering, Institute of Electronics and Telecommunications, Poznań University of Technology, Poland, Sep. 2006.

17. M Wódczak. Aspects of Cross-Layer Design in Autonomic Cooperative Networking. *IEEE Third International Workshop on Cross Layer Design, Rennes, France*, 30 November - 1 December 2011.
18. M Wódczak. Autonomic Cooperation in Ad-hoc Environments. *5th International Workshop on Localised Algorithms and Protocols for Wireless Sensor Networks (LOCALGOS) in conjunction with IEEE International Conference on Distributed Computing in Sensor Systems (DCOSS), Barcelona, Spain*, 27-29 June 2011.
19. M. Wódczak. Autonomic Cooperative Networking for Wireless Green sensor Systems. *International Journal of Sensor Networks (IJSNet)*, 10(1/2), 2011.
20. M Wódczak. Convergence Aspects of Autonomic Cooperative Networking. *IEEE Fifth International Conference on Next Generation Mobile Applications, Services and Technologies, Cardiff, Wales, UK*, 14-16 September 2011.
21. M Wódczak. Resilience Aspects of Autonomic Cooperative Communications in Context of Cloud Networking. *IEEE First Symposium on Network Cloud Computing and Applications, Toulouse, France*, 21-23 November 2011.
22. M. Wódczak. Cooperative Re-Routing. *full patent application no. 13/042701 filed to the United States Patent and Trademark Office*, 8 March 2011.
23. M. Wódczak, T. B. Meriem, R. Chaparadza, K. Quinn, B. Lee, L. Ciavaglia, K. Tsagkaris, S. Szott, A. Zafeiropoulos, B. Radier, J. Kielthy, A. Liakopoulos, A. Kousaridas, and M. Duault. Standardising a Reference Model and Autonomic Network Architectures for the Self-managing Future Internet. *IEEE Network*, 25(6):50–56, November/December 2011.
24. Y. R. Zheng and C. Xiao. Simulation Models with Correct Statistical Properties for Rayleigh Fading Channels. *IEEE Transactions on Communications*, 51(6):920–928, Jun. 2003.

Chapter 6
Cooperative Autonomic Network Deployments

6.1 Introduction

Following the extensions to the Generic Autonomic Network Architecture proposed in the previous chapter and aiming to incorporate the Routing information Enhanced Algorithm for Cooperative Transmission, aided with Virtual Antenna Arrays, into the global picture of the autonomic system design, this chapter looks into applying them to support autonomic cooperative networking in Relay Enhanced Cell. To this end, a specific local area indoor scenario is investigated, where radio propagation is severely distorted because of numerous obstacles in the form of walls. The evaluations are performed with the use Distributed Space-Time Block Coding for a number of different deployments of Fixed Relay Nodes in order to understand the capabilities of an equivalent autonomic system, switching among FRNs belonging to a mesh of Radio Access Points to meet the Quality of Service requirements. This analysis will be further extended to Emergency Communications in the next chapter.

6.2 Relay Enhanced Cell

The indoor local area scenario chosen for evaluation allows to present the potential of an autonomic system [14], [2], which would be able to choose the cooperating FRNs out of a fixed mesh of Radio Access Points (RAPs). This scenario is characterised by a considerable user density, high shadowing and heavy signal attenuation due to the existence of obstacles represented by numerous walls [7]. Thanks to its isolated characteristic, the scenario offers advantages such as low interference when compared to the outdoor cases. As shown in Figure 6.1, it consists of one floor of a height of 3 m in a building, where two corridors of dimensions of 5 m x 100 m and 40 rooms of dimensions of 10 m x 10 m are located. Typically, there are four Fixed Relay Nodes (FRNs) put in the corridors and the transmission is coordinated by the BS placed in the centre. The base deployment assumes the FRNs to be placed

in the middle of the corridors, 25 m and 75 m away from either the left or right side
of the building. The analysis is performed according to the general assumptions for

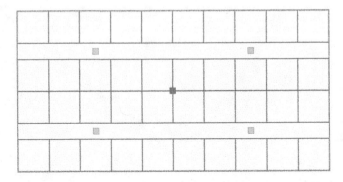

Fig. 6.1 Baseline relay deployment for the indoor scenario

the indoor environment provided in [7]. Besides, specific additional parameters are
taken into account as outlined below. In particular, as the scenario under investigation is symmetric, the simulation area may be limited to a set of 10 rooms located
in the right bottom part of Figure 6.1, next to the FRN. For this reason the region
is presented and referred to in the figures and descriptions provided in this chapter.
Additionally, fixed modulation and coding scheme is employed consisting of QPSK
modulation and (4, 5, 7) convolutional code. AWGN channel is assumed together
with the A1 radio propagation model [7], not only for the links between FRNs and
UTs, but also between BS and FRNs. Depending on the presence of walls, either its
Line-of-Sight (LOS) or Non Line-of-Sight (NLOS) version is employed. The LOS
model is defined in the following way:

$$PL_{LOS}[dB] = 18.7\log_{10}(d) + 46.8 + \sigma \tag{6.1}$$

where d denotes the distance in meters between the transmitter and the receiver and
σ represents the standard deviation of shadow fading and is equal to 3 dB. In turn,
the NLOS propagation model is described as:

$$PL_{NLOS}[dB] = 20.0\log_{10}(d) + 46.4 + 5n_w + \sigma \tag{6.2}$$

where n_w denotes the number of walls between the transmitter and the receiver and
σ is equal to 6dB. This means that all the walls are assumed to be of the same light
type.

More detailed parameters are given in Table 6.1. In particular, Orthogonal Frequency Division Multiple Access (OFDMA) is employed and Time Division Duplex (TDD) mode used at the carrier frequency of 5.0 GHz, assuming channel bandwidth of 100 MHz. The transmission power for BS, FRN and UT is at the level of 21
dBm, whereas the antenna gains are equal to 14 dBi, 7 dBi and 0 dBi, respectively.

Table 6.1 System parameters

Parameter	Value	Comments
Carrier frequency	5.0 GHz	TDD mode
Channel bandwidth	100 MHz	OFDMA
Spatial processing	Distributed STBC	FRN-FRN cooperation
BS antennas	1	Omnidirectional
FRN antennas	1	Omnidirectional
UT antennas	1	Omnidirectional
BS transmit power	21 dBm	14 dBi antenna gain
FRN transmit power	21 dBm	7 dBi antenna gain
UT transmit power	21 dBm	0 dBi antenna gain
Channel modelling	AWGN channel	A1 NLOS Room-Room model used for both BS-FRN and FRN-UT links (also for Room-Corridor transmission)
Link adaptation	Fixed code and modulation scheme	QPSK and (4, 5, 7) convolutional code
Mobility	Yes	User terminals
Resource scheduling	Fixed	Each user was assigned 1 chunk (8 subcarriers and 15 OFDM symbols)
RAP selection	Signal power	At the destination
Traffic model	Constant bit rate	CBR

The noise figure at the receiver is equal to 7 dB and the noise power spectral density amounts to -174 dBm/Hz. For the baseline deployment presented in Figure 6.1, each user is assigned one chunk of the radio resources, spreading over 8 subcarriers and 15 OFDM symbols. The average interference power level per subcarrier is at the level of -125 dBm.

6.3 Radio Resource Partitioning

The structure of the super-frame for the base deployment is defined according to [7] and radio resources are partitioned in both the temporal and spectral domains (Figure 6.2) [6]. The spatial domain is also exploited and it is visible in the cases of

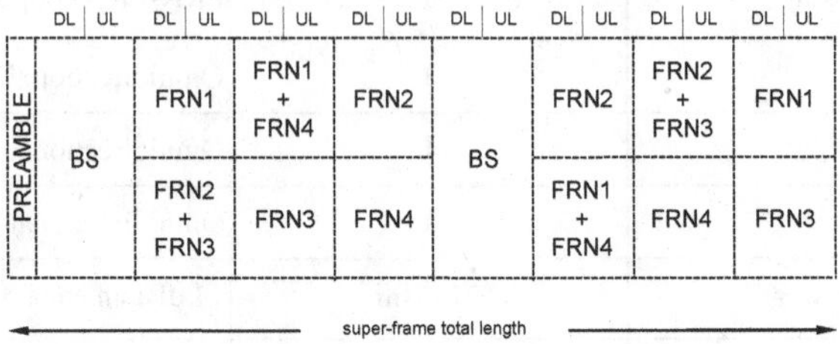

Fig. 6.2 Radio resource partitioning

cooperation between FRNs. Specifically, following the preamble, a very similar pattern is repeated twice during one super-frame [7]. First, the resources are assigned to the BS and then to different combinations of FRNs. Consequently, three FRNs may be active simultaneously, when two of them operate cooperatively in the spatio-temporal mode [5], [13]. The simulation results presented in the remainder of the chapter pertain to the following three cases: direct transmission, single path (conventional) relaying and Fixed Relay Node to Fixed Relay Node (FRN-FRN) cooperation (cooperative relaying). In the latter case, pairs of selected FRNs form Virtual Antenna Arrays [4] and perform the operation of a Distributed Space-Time Block Coding [8], [12]. The size of VAA is assumed to be equal to two nodes and so the G_2 code is exploited [1], [11].

In particular, in all cases, the attainable QoS is evaluated from the perspective of the relative user throughput [13], [17]. Such throughput is calculated as the ratio between the number of bits transmitted successfully and the total number of bits sent. All the presented results are obtained assuming that a given FRN takes part in cooperation even if decoding has been unsuccessful. In this way the degradation in performance introduced by a wrongly positioned FRN is augmented. Obviously, it should be read as an indication that such positioning is not of interest and the FRN would have to remain silent in normal system operation. The reference results obtained for such a non-autonomic system basically show that it is possible to make up for the performance degradation visible in Figure 6.3(b) with the aid of single path relaying, as shown in Figure 6.3(c). Unfortunately, the gain provided by the FRN-FRN cooperation in Figure 6.3(d) seems almost diminishable when compared to the single path relaying case. The reason is that the signal coming from the cooperating FRN2 to the destination UT is usually heavily attenuated by a higher number of walls. According to the aforementioned A1 NLOS propagation model [7], each wall between rooms attenuates the transmitted signal by 5 dB. As a result, the power levels of the signals received by the destination UT from both the cooperating FRNs

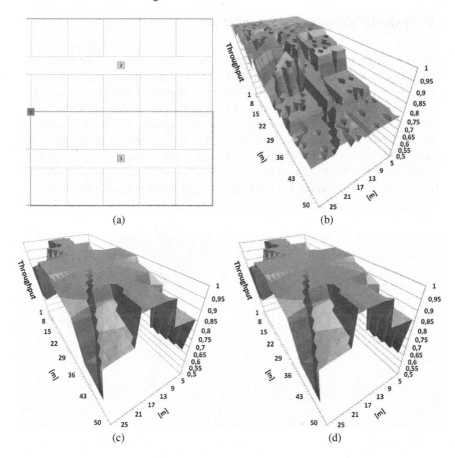

Fig. 6.3 Deployment of FRNs (a) and relative user throughput for direct transmission (b), single path relaying via FRN3 (c) and FRN2-FRN3 cooperation (d)

may differ, for example, even up to about 15 dB. It means that quite often, the signal coming from the other Fixed Relay Node, as it is the case for FRN2 in Figure 6.3(a), is too weak to constructively be added to improve throughput. Consequently, in such cases, there might be no gain visible from distributed spatio-temporal processing performed by the cooperating FRNs. This is one of the main reasons for including the notion of autonomic behaviours, so the system would be able to self-configure, especially if there is a bigger mesh of FRNs available to choose from, as discussed in the Section 6.4. This approach will be also further extended in Chapter 7 [18], [20] to the Mobile Relay Node (MRN) case for autonomic cooperative networking in emergency communications systems [16], [15].

6.4 Towards Autonomic Cooperation

In order to verify potential gains resulting from the employment of the Autonomic
Cooperative System Design principles [3], [21] to Relay Enhanced Cell [7], several different FRN set-ups are verified. This is to investigate whether an autonomic
system, able to monitor the network and then apply policies, would profit from a
more sophisticated deployment of FRNs resulting from the possibility of choosing
the best FRNs out of a mesh of RAPs [15], [17]. For the following evaluation figures, there is always the FRN2-FRN3 cooperation performance shown in part (b),
whereas the relative throughputs achievable by single path relaying via FRN2 and
FRN3 are shown in parts (c) and (d), respectively. First, it is assumed that the deci-

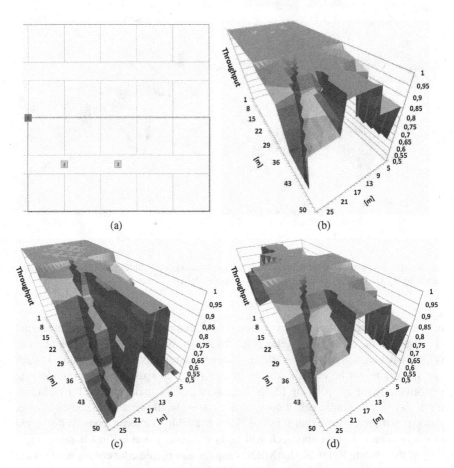

Fig. 6.4 Deployment of FRNs (a) and relative user throughput for FRN2-FRN3 cooperation (b),
single path relaying via FRN2 (c) and single path relaying via FRN3 (d)

sion entities would choose to keep using the Autonomic Cooperative Node (ACN) denoted as FRN3 and located at its previous position while selecting the other ACN, denoted as FRN2 and located also in the area of interest. In the case of Figure 6.4(a), the FRN2 is placed closer to the BS so improvement is observable. Although, compared to the reference case, the throughput attainable thanks to FRN-FRN cooperation is higher in some regions, both FRNs could be also exploited separately to cover different parts of the region of interest. In this way the single path relaying via FRN2 would provide even higher throughput in the rooms located in the immediate vicinity of the BS and in the corridor, as depicted in Figure 6.4(c), while FRN3 could serve the users located in the remaining area, according to Figure 6.4(d). Following,

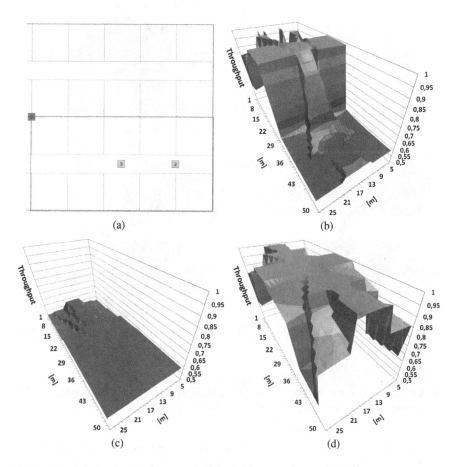

Fig. 6.5 Deployment of FRNs (a) and relative user throughput for FRN2-FRN3 cooperation (b), single path relaying via FRN2 (c) and single path relaying via FRN3 (d)

the Decision Entities would keep FRN3 at its position and chose as FRN2 the ACN placed farther from the BS, to verify whether it is possible to compensate for the throughput degradation visible in the distant corners of the investigated scenario.

Unfortunately, according to Figure 6.5, the results seem worse because the distance is significant. And so, in spite of reasonable performance of the FRN3, visible in Figure 6.5(d), the FRN-FRN cooperation would be negatively affected and inapplicable, as it is proved by Figure 6.5(c). Keeping in mind that in the aforementioned

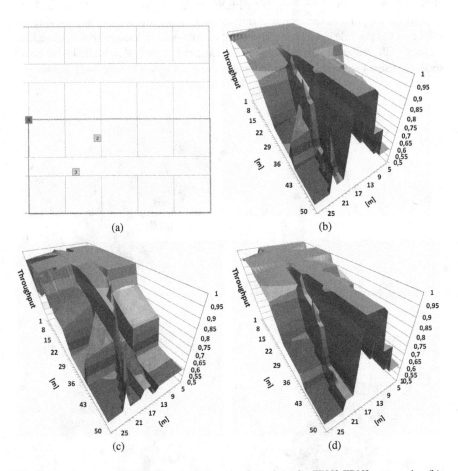

Fig. 6.6 Deployment of FRNs (a) and relative user throughput for FRN2-FRN3 cooperation (b), single path relaying via FRN2 (c) and single path relaying via FRN3 (d)

two cases the distance and therefore the number of obstacles between the BS and each of the FRNs is different, two other deployments are evaluated as shown in Figure 6.6 and Figure 6.7. This time it is assumed that the Autonomic Cooperative Entities of the GANA architecture would make sure that the number of walls in-between the aforementioned BS-FRN pairs is kept the same and the distances are similar. The throughput achievable in the vicinity of the BS (Figure 6.6) seems rather reasonable but the rest of the area is still covered insufficiently.

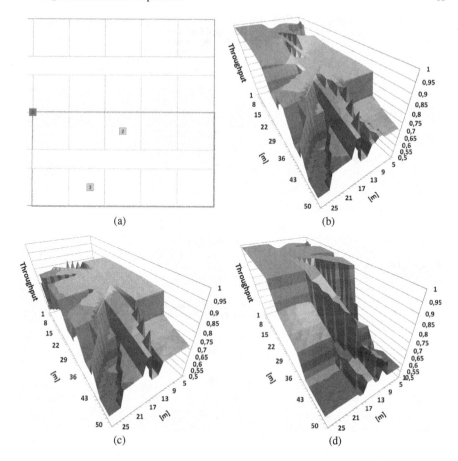

Fig. 6.7 Deployment of FRNs (a) and relative user throughput for FRN2-FRN3 cooperation (b), single path relaying via FRN2 (c) and single path relaying via FRN3 (d)

To further understand the investigated indoor scenario, selected deployments are analysed from the perspective of the Cumulative Distribution Function (CDF) in relation to the reference one for the BS only scenario (central node in Figure 6.1). The CDF for the deployment presented in Figure 6.4 is shown in Figure 6.8 and it proves that such a system tends to behave better when compared to the reference scenario, however, there appear low relative throughput values, too. Moving one of the FRNs even farther apart as in Figure 6.5 clearly shows that the influence of walls is really significant (Figure 6.9). The obtained results are even worse than for the reference scenario pointing significant drop in the average relative throughput. Finally, when FRNs are deployed in the manner outlined in Figure 6.6, the performance gets improved again but the first investigated case still offers much better CDF.

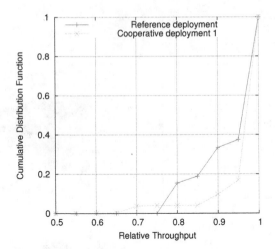

Fig. 6.8 CDF for scenario No 1

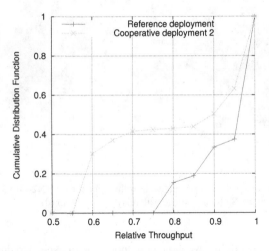

Fig. 6.9 CDF for scenario No 2

6.5 Advanced Cooperative Deployments

The main conclusion from the presented considerations is that the scenario is very specific, even if the autonomic cooperative system design principles were employed. Certain compromise deployment, potentially meeting the criteria related to policies, monitoring and user requirements, not necessarily needs to form the optimum solution for every case. In other words, one could expect that the autonomic routines driving the network of the future should be in a position to go one step further and be able to change the classification of nodes from FRNs to additional BSs, when

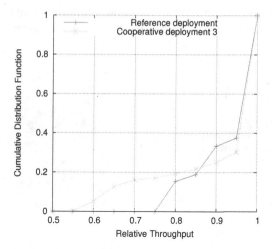

Fig. 6.10 CDF for scenario No 3

necessary. Such a change would allow FRNs to coordinate transmission and not just relay the data. Obviously, this approach would demand certain dose of cognition as the system would have to make decisions going beyond the simple notion of being autonomic [10]. The analysed scenario is a perfect example of such a situation, where additional gains could be achieved thanks to advanced modifications to the deployment of Radio Access Points, meaning that, for example, a node primarily designed to act as FRN could take the role equivalent to that of a BS. An exam-

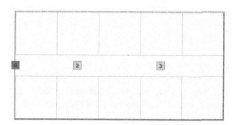

Fig. 6.11 Modified deployment

ple set-up of this type is depicted in Figure 6.11. Clearly, certain advancement in equipment design would be required so the devices were able to expose capabilities allowing the autonomic network to configure them as needed, for example with the use of accordingly defined profiles [17], [20]. As outlined in [9], cloud networking may be actually used for decoupling radio heads from the Base Stations, so the relevant BS equipment may be located in data centres [19]. It means that upgrading BS systems to various new standards may be completed with software upgrades only, without even touching the hardware [9]. In other words, BS may be perceived as a universal Radio Access Point (RAP) and so it could be downgraded to FRN and vice versa. The simulation results provided in Figure 6.12 pertain to the set-up outlined

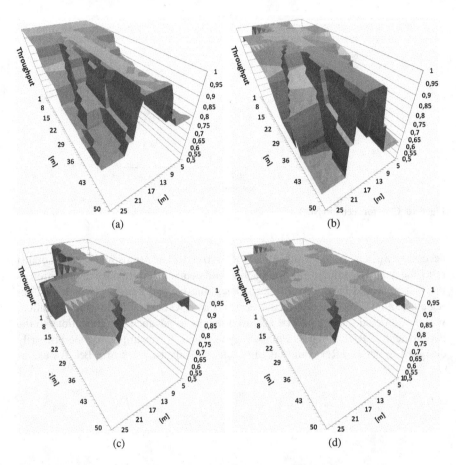

Fig. 6.12 Relative user throughput in the modified deployment for direct transmission (a), single path relaying via FRN2 (b), single path relaying via FRN3 (c) and FRN2-FRN3 cooperation (d)

in Figure 6.11. In fact, the baseline case (Figure 6.1) poses difficulties in terms of showing the full gains achievable from the FRN-FRN cooperation. This is because deploying FRNs in such a way that the obstacles are somehow omitted seems very difficult, if not impossible. In the new approach, the autonomic cooperative network would be able to deploy additional BS so that two of them could serve the whole area jointly, each placed in the middle of the corridor it is positioned in. Based on the conclusions from the previously presented results, each BS would be assisted by four FRNs, two on its left and two on its right side. Analysing the outcome, one can conclude that this time the FRN-FRN cooperation performs in the best possible way. The reason is that the BS is able to provide maximised throughput almost in the whole corridor and, as a result, both cooperating FRNs are fed with radio signal of a very good quality. Consequently, they can retransmit the received packets cooperatively and provide better coverage as compared to the case when the BS is switching between two FRNs working in the conventional mode. In practice, one could make use of the fact that the BSs located in different corridors are separated by a number of quite heavy walls. Keeping in mind that these walls were successfully spoiling the FRN-FRN cooperation, this time they would obviously help to suppress interference.

6.6 Conclusion

In this chapter, the notion of autonomic cooperative network deployments was introduced on the basis of the previously proposed extensions to the Generic Autonomic Network Architecture. In particular, the Distributed Space-Time Block Coding was exploited for the purposes of quantifying the throughput gains potentially achievable thanks to different positioning of Fixed Relay Nodes, forming Virtual Antenna Arrays. In this way it was possible to analyse the capabilities of an equivalent system which would be able to select FRNs from a mesh of Radio Access Points.

References

1. S. Alamouti. A Simple Transmit Diversity Technique for Wireless Communications. *IEEE Journal on Selected Areas in Communications*, 16(8):1451–1458, Oct. 1998.
2. R. Chaparadza, L. Ciavaglia, M. Wódczak, C.-C. Chen, B.A. Lee, A. Liakopoulos, A. Zafeiropoulos, E. Mancini, U. Mulligan, A. Davy, K. Quinn, B. Radier, N. Alonistioti, A. Kousaridas, P. Demestichas, K. Tsagkaris, M. Vigoureux, L. Vreck, M. Wilson, and L. Ladid. ETSI Industry Specification Group on Autonomic network engineering for self-managing Future Internet (ETSI ISG AFI). *10th International Conference on Web Information Systems Engineering, Poznań, Poland*, Sep. 2009. Published in Springer Lecture Notes in Computer Science (LNCS): Web Information Systems Engineering, Vol. 5802/2009, edited by G. Vossen, D. Long, and J. Yu.
3. R. Chaparadza, S. Papavassiliou, T. Kastrinogiannis, M. Vigoureux, E. Dotaro, K. A. Davy, M. Quinn, Wódczak, and A. Toth. Creating a viable Evolution Path towards Self-Managing Future Internet via a Standardizable Reference Model for Autonomic Network Engineering. *Chapter in the book: "Towards the Future Internet - A European Research Perspective" edited by G. Tselentis, J. Domingue, A. Galis, A. Gavras, D. Hausheer, S. Krco, V. Lotz, and T. Zahariadis, published by IOS Press, ISBN: 978-1-60750-007-0*, May 2009. Also published at the Future Internet Assembly 2009 in Prague.
4. M. Dohler, A. Gkelias, and H. Aghvami. A resource allocation strategy for distributed MIMO multi-hop communication systems. *IEEE Communications Letters*, 8(2):99–101, Feb. 2004.
5. K. Doppler, A. Osseiran, M. Wódczak, and P. Rost. On the Integration of Cooperative Relaying into the WINNER System Concept. *16th IST Mobile & Wireless Communications Summit 2007, Budapest, Hungary*, 1-5 July 2007.
6. K. Doppler, S. Redana, M. Wódczak, P. Rost, and R. Wichman. Dynamic resource assignment and cooperative relaying in cellular networks: Concept and performance assessment. *EURASIP Journal on Wireless Communications and Networking*, Jul. 2007.

7. M. Dottling, R. Irmer, K. Kalliojarvi, and S. Rouquette-Leveil. System Model, Test Scenarios, and Performance Evaluation. *Chapter in the book: "Radio Technologies and Concepts for IMT-Advanced", edited by M. Dottling, W. Mohr and A. Osseiran, published by Wiley, ISBN: 978-0-470-74763-6*, December 2009.

8. J. N. Laneman and G. W. Wornell. Distributed space-time-coded protocols for exploiting cooperative diversity in wireless networks. *IEEE Transactions on Information Theory*, 49(10):2415–2425, Oct. 2003.

9. Y. Lin, L. Shao, Z. Zhu, Q. Wang, and R. K. Sabhikhi. Wireless network cloud: Architecture and system requirements. *IBM Journal of Research and Development*, 54(1):4:1 – 4:12, Jan.-Feb. 2010.

10. M. Smirnov, J. Tiemann, R. Chaparadza, Y. Rebahi, S. Papavassiliou, V. Karyotis, V. Pouli, V. Merekoulias, A. Gulyas, Z. Heszberger, F. Retvari, G. Nemeth, M. Wódczak, V. Kaldanis, A. Markopoulos, G. Karantonis, A. Davy, K. Quinn, S. Cheng, Y. Li, Y. Jin, X. Gong, Y. Cui, B. Hu, Y. Shi, W. Wang, A. Liakopoulos, J. Zafeiropoulos, J. Lopez, J. Munoz, M. Vigoreux, B. Berde, D. Cleary, and A. Toth. Demystifying Self-awareness of Autonomic Systems. *ICT Mobile Summit, Santander, Spain*, 10 - 12 June 2009.

11. M. Wódczak. *On Routing information Enhanced Algorithm for space-time coded Cooperative Transmission in wireless mobile networks*. PhD thesis, Faculty of Electrical Engineering, Institute of Electronics and Telecommunications, Poznań University of Technology, Poland, Sep. 2006.

12. M. Wódczak. Extended REACT - Routing information Enhanced Algorithm for Cooperative Transmission. *16th IST Mobile & Wireless Communications Summit 2007, Budapest, Hungary*, 1-5 July 2007.

13. M. Wódczak. Cooperative Relaying in an Indoor Environment. *ICT Mobile Summit, Stockholm, Sweden*, 10-12 June 2008.

14. M. Wódczak. Future Autonomic Cooperative Networks. *Second International ICST Conference on Mobile Networks and Management, Santander, Spain*, Sep. 2010. Published in Springer Lecture Notes of the Institute for Computer Sciences, Social Informatics and Telecommunications Engineering (LNICST): Mobile Networks and Management, Vol. 68/2011, edited by K. Pentikousis, R. Aguero, M. Garcia-Arranz, and S. Papavassiliou.

15. M Wódczak. Aspects of Cross-Layer Design in Autonomic Cooperative Networking. *IEEE Third International Workshop on Cross Layer Design, Rennes, France*, 30 November - 1 December 2011.

16. M Wódczak. Autonomic Cooperation in Ad-hoc Environments. *5th International Workshop on Localised Algorithms and Protocols for Wireless Sensor Networks (LOCALGOS) in conjunction with IEEE International Conference on Distributed Computing in Sensor Systems (DCOSS), Barcelona, Spain*, 27-29 June 2011.

17. M Wódczak. Convergence Aspects of Autonomic Cooperative Networking. *IEEE Fifth International Conference on Next Generation Mobile Applications, Services and Technologies, Cardiff, Wales, UK*, 14-16 September 2011.

18. M. Wódczak. Deployment Aspects of Autonomic Cooperative Communications in Emergency Networks. *3rd International Congress on Ultra Modern Telecommunications and Control Systems, IEEE ICUMT, Budapest, Hungary*, 5-7 October 2011.

19. M Wódczak. Resilience Aspects of Autonomic Cooperative Communications in Context of Cloud Networking. *IEEE First Symposium on Network Cloud Computing and Applications, Toulouse, France*, 21-23 November 2011.

20. M. Wódczak. Convergence Aspects of Autonomic Cooperative Networks. *International Journal of Information Technology and Web Engineering (IJITWE)*, 2012. Accepted for publication.

21. M. Wódczak, T. B. Meriem, R. Chaparadza, K. Quinn, B. Lee, L. Ciavaglia, K. Tsagkaris, S. Szott, A. Zafeiropoulos, B. Radier, J. Kielthy, A. Liakopoulos, A. Kousaridas, and M. Duault. Standardising a Reference Model and Autonomic Network Architectures for the Self-managing Future Internet. *IEEE Network*, 25(6):50–56, November/December 2011.

Chapter 7
Autonomic Emergency Communications

7.1 Introduction

Emergency networks formed by First Responders, or in fact by the communications equipment they carry, seem to form a very relevant area for the application of the concept of autonomic cooperative network deployments, as introduced in the previous chapter. This holds true especially because of the ad hoc and/or mesh nature of such network set-ups, where one may observe very dynamic topology changes. Consequently, a truly mobile autonomic cooperative relaying might be possible as opposed to the need for deploying a mesh of fixed Relay Nodes. In particular, the emergency system design is discussed from the perspective of enabling the Virtual Antenna Array aided cooperative transmission in such an environment. To this end, the Equivalent Distributed Space-Time Block Encoder is used for general evaluations assuming an external Base Station, provided by the Mobile Emergency Operations Centre, as well as dynamic Radio Access Point selection is applied.

7.2 Emergency System Design

The currently investigated advancements in the development of novel communications infrastructure for the emergency system of the future seem very demanding in terms of the technologies to be applied [7], [3]. It is especially visible in the ad hoc and/or mesh part of the network, where devices carried by First Responders (FRs) seek seamless and on-demand connectivity for the transmission of different Quality of Service requirements [13], [11]. For this very reason, emergency networks formed by FRs operating in the area of incident seem to have become a very relevant field for the application of the previously introduced concepts related to the autonomic cooperative system design, as discussed in Chapter 5 and Chapter 6. It especially holds true for numerous small groups of FRs, coordinated by their respective Chief First Responders (CFRs). Such groups may be assumed to contain from 4 up

to 6 FRs, which immediately translates into certain possibilities when the preferred
network topology is concerned [15]. In other words, it might mean that multi-hop

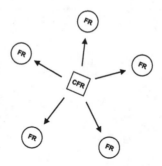

Fig. 7.1 Baseline set-up

communications between the CFR and its FRs might not be the predominant case
because FRs would normally gather around the CFR and form the topology of a star,
as depicted in Figure 7.1. On the other hand, multi-hop communications cannot be

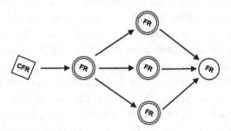

Fig. 7.2 Multi-hop set-up

obviously excluded, especially when such a group is more spread apart, as indicated
in Figure 7.2. Moreover, the option of having bigger or merging and splitting FR
groups might also need to be taken into account. This brings about the question of
how the transmission should be organised so the system would be highly resilient
to the harsh environment in which it is intended to operate [16]. In principle, the
implementation of the autonomic system concept, supported by the inclusion of cooperative behaviours, seems to be the proper direction [12]. This is, in fact, already
visible in Figure 7.2, where a group of FRs may form a Virtual Antenna Array [5]
in an autonomic way [15]. As a result, both the idea behind Autonomic Cooperative
Transmission (ACT) (Section 5.5) and Autonomic Cooperative Re-Routing (ACRR)
(Section 5.6) would immediately become applicable also here.

Besides, cooperation may also have another dimension in the investigated case.
The reason is that, from the emergency system perspective, apart from CFRs and
FRs there also exist both the Emergency Operations Centre (EOC) and the Mobile

Emergency Operations Centre (MEOC). The former is situated at a fixed position, while the latter is mobile and always relocated to the area of incident. Consequently, the following two cases of cooperation between or among CFRs need to be taken into consideration [15]. First of all, it is typically assumed that the process of communication between two FRs belonging to two separate FR groups would be assisted by their respective CFRs, communicating not directly but via MEOC, as outlined in Figure 7.3. This assumption would need to be relaxed because of potential lack of

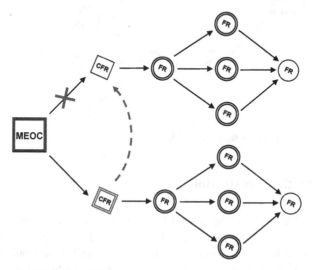

Fig. 7.3 Communication between CFRs

communication with MEOC, which would be a presumably rare case [15]. Moreover, one would need to consider the possibility that in the initial phase, when FR groups are already in the field, the MEOC could be still on its way. Secondly, for legacy reasons, related to the hierarchy of a consolidated management of distinct FR groups, it would be required that only a given CFR can communicate and so route external data stream(s), coming from the EOC, towards a given FR team. From the system design perspective, there are no clear arguments, however, against the establishment of a logical link via an external CFR, as long as it is transparent to the system so the hierarchy would look as if nothing had changed. In fact, in the example outlined in Figure 7.4, one of the CFRs is not able to communicate with its FRs. The system would support such a case through autonomic switching to a back-up mode, where another CFR could be exploited as an intermediary entity, allowing for communication with MEOC.

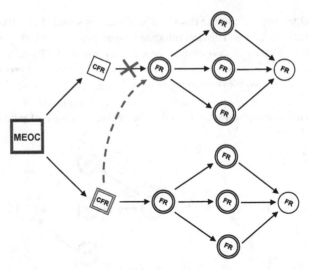

Fig. 7.4 Back-up operation of supporting CFR

7.3 Autonomic Cooperation

In order to enable switching between different operation modes, one could employ certain entities allowing for autonomic decisions, as already discussed in previous chapters [19], [4]. In this way not only the links between CFRs and their corresponding FRs could be monitored but the system would be also notified when another CFR could potentially serve the FR(s) not belonging to his own team [2]. This way better robustness could be offered and the generic operations performed by MEOC and FR could be summarised, as outlined in Figure 7.5. Having the relevant global data related to the network parameters, MEOC would even have some leeway to arrange for cooperation still before link degradation occurs. As an example, the assistance of an external CFR in the communication between MEOC and a given FR, done by means of cooperative transmission, and again in a way transparent to the system and not affecting the hierarchy, is presented in Figure 7.6. Given the fact that a specific FR might become exposed to severe radio channel impairments, resulting either from obstacles or too big a distance towards their CFR, another CFR may support the communication so that diversity gain offered by virtual MISO channel can be exploited. The destination FR is served cooperatively by both the CFRs. Obviously, this would require the application of the Distributed Space-Time Block Coding [8], so that both CFR may logically form a Virtual Antenna Array and apply physical layer signal processing techniques to orthogonalise the wireless radio channel [5], [1]. Consequently, appropriate data would have to be delivered from MEOC and that the system should be properly synchronised.

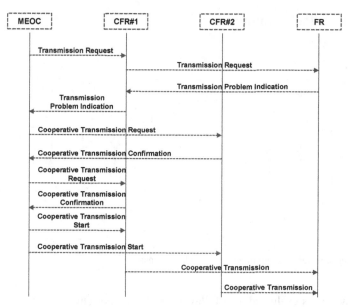

Fig. 7.5 Interactions between MEOC and FR sides

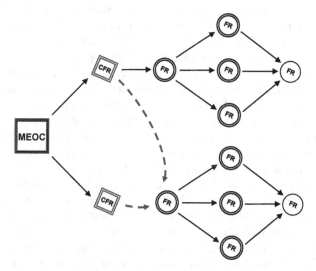

Fig. 7.6 Cooperative transmission supported by the other CFR

7.4 System Performance

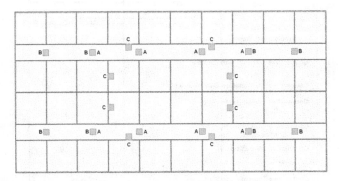

Fig. 7.7 Baseline relay deployment for the indoor scenario

The system is evaluated as a follow-up to the indoor environment (Figure 6.1) introduced in Chapter 6. Generally, the parameters given therein (Table 6.1) are applied [6], [10]. The exception is that the Base Station is not placed inside the area of interest. Instead there is MEOC located outside, which is equipped with a Base Station capable of providing a much higher dynamic transmit power range. Consequently, an assumption is made that the first hop links could be considered lossless and so the Equivalent Distributed Space-Time Block Encoder is applied. The function of RNs is assumed by CFRs who, on the one hand, communicate with MEOC and, on the other hand, form VAAs and feed data cooperatively towards FRs. This way the influence of the positioning of CFRs on the throughput, which can be supported for FRs located at different positions, is investigated. Additionally, the switching among pairs of best CFRs for a given deployment is done autonomically and in a dynamic way, on the basis of the signal strength observed by the destination FR(s). Consequently, the CFRs are assumed to follow the positioning pattern A, B, or C, as outlined in Figure 7.7. The corresponding results are presented in Figure 7.8(a), Figure 7.8(b), and Figure 7.8(c), respectively. Analysing the results and comparing them with Chapter 6, one may conclude that in most of the cases it is possible to obtain full throughput. Moreover, throughput gain and loss between pairs of the investigated deployments is compared. In particular, Figure 7.9 outlines the throughput gain between Deployments A and B where in the central part there is possible to gain up to about 60 per cent, but there is also a gradual loss towards the corners with a maximum drop of about 40 per cent. Following, Figure 7.10 outlines throughput gain and loss between Deployments B and C. There is loss of about up to 40 per cent in the centre, while the distant corners gain up to 80 percent. Finally, Figure 7.10 outlines throughput gain and loss between Deployments C and A. In this case, obviously Deployment C is generally outperformed by Deployment A, as clearly visible in the distant parts of the scenario. The presented results prove that

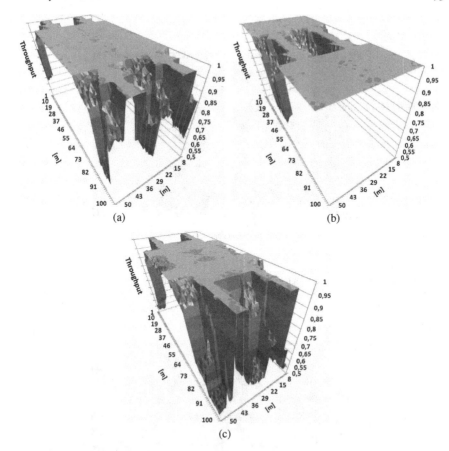

Fig. 7.8 Relative throughput for: (a) Deployment A, (a) Deployment B, (c) Deployment C

there is no single optimum deployment and autonomic switching for more dynamic set-ups would be necessary.

7.5 Way Forward

The option of investigating the concept of autonomic cooperative networking in the context of emergency communications is very appealing because the limitation, related to the need for deploying a mesh of Fixed Relay Nodes to select from, as visible in the previous chapter, does not apply here. In this way a possibility for additional research in the area of a convergent solution is envisaged [14], [17]. In particular, one might look into Relay Enhanced Cell and the employment of Mobile Relay Nodes represented by Chief First Responders, so the concept of REACT

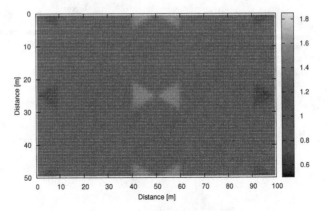

Fig. 7.9 Throughput gain and loss between Deployments A and B

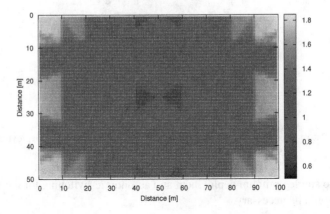

Fig. 7.10 Throughput gain and loss between Deployments B and C

[9], [8] could be used in parallel for a number of ad hoc FR teams. This would also involve the previously introduced extensions to GANA in the form of the Decision Elements orchestrating Cooperative Transmission and Cooperative Re-Routing [18].

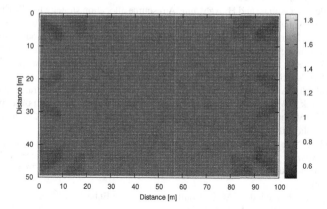

Fig. 7.11 Throughput gain and loss between Deployments C and A

7.6 Conclusion

In this chapter, the idea of autonomic cooperative network deployments was extended to emergency communications systems. Such systems are naturally characterised by ad hoc nature of the topology formed by Chief First Responders and First Responders and so there is no need for deploying any fixed mesh of Relay Nodes. In particular, the performance of the Equivalent Distributed Space-Time Block Encoder was evaluated for selected configurations of the cooperating CFRs to quantify the achievable gains.

References

1. S. Alamouti. A Simple Transmit Diversity Technique for Wireless Communications. *IEEE Journal on Selected Areas in Communications*, 16(8):1451–1458, Oct. 1998.
2. A.Liakopoulos, A.Zafeiropoulos, A.Polyrakis, M.Grammatikou, J.M.Gonzalez, M. Wódczak, and R.Chaparadza. Monitoring Issues for Autonomic Networks: The EFIPSANS Vision. *European Workshop on Mechanisms for the Future Internet*, 2008.
3. G. Calarco, M. Casoni, A. Paganelli, D. Vassiliadis, and M Wódczak. A Satellite based System for Managing Crisis Scenarios: the E-SPONDER Perspective. *5th Advanced Satellite Multimedia Systems Conference, Cagliari, Italy*, 13-15 September 2010.
4. R. Chaparadza, L. Ciavaglia, M. Wódczak, C.-C. Chen, B.A. Lee, A. Liakopoulos, A. Zafeiropoulos, E. Mancini, U. Mulligan, A. Davy, K. Quinn, B. Radier, N. Alonistioti, A. Kousaridas, P. Demestichas, K. Tsagkaris, M. Vigoureux, L. Vreck, M. Wilson, and L. Ladid. ETSI Industry Specification Group on Autonomic network engineering for self-managing Future Internet (ETSI ISG AFI). *10th International Conference on Web Information Systems Engineering, Poznań, Poland*, Sep. 2009. Published in Springer Lecture Notes in Com-

puter Science (LNCS): Web Information Systems Engineering, Vol. 5802/2009, edited by G. Vossen, D. Long, and J. Yu.

5. M. Dohler, A. Gkelias, and H. Aghvami. A resource allocation strategy for distributed MIMO multi-hop communication systems. *IEEE Communications Letters*, 8(2):99–101, Feb. 2004.

6. M. Dottling, R. Irmer, K. Kalliojarvi, and S. Rouquette-Leveil. System Model, Test Scenarios, and Performance Evaluation. *Chapter in the book: "Radio Technologies and Concepts for IMT-Advanced", edited by M. Dottling, W. Mohr and A. Osseiran, published by Wiley, ISBN: 978-0-470-74763-6*, December 2009.

7. D. Vassiliadis, A. Garbi, G. Calarco, M. Casoni, A. Paganelli, R. Morera, C.-M. Chen, and M Wódczak. Wireless Networks at the Service of effective First Response Work: the E-SPONDER Vision. *EEE International Symposium on Wireless Pervasive Computing, Modena, Italy*, 5-7 May 2010.

8. M. Wódczak. *On Routing information Enhanced Algorithm for space-time coded Cooperative Transmission in wireless mobile networks*. PhD thesis, Faculty of Electrical Engineering, Institute of Electronics and Telecommunications, Poznań University of Technology, Poland, Sep. 2006.

9. M. Wódczak. Extended REACT - Routing information Enhanced Algorithm for Cooperative Transmission. *16th IST Mobile & Wireless Communications Summit 2007, Budapest, Hungary*, 1-5 July 2007.

10. M. Wódczak. Cooperative Relaying in an Indoor Environment. *ICT Mobile Summit, Stockholm, Sweden*, 10-12 June 2008.

11. M. Wódczak. Future Autonomic Cooperative Networks. *Second International ICST Conference on Mobile Networks and Management, Santander, Spain*, Sep. 2010. Published in Springer Lecture Notes of the Institute for Computer Sciences, Social Informatics and Telecommunications Engineering (LNICST): Mobile Networks and Management, Vol. 68/2011, edited by K. Pentikousis, R. Aguero, M. Garcia-Arranz, and S. Papavassiliou.

12. M Wódczak. Aspects of Cross-Layer Design in Autonomic Cooperative Networking. *IEEE Third International Workshop on Cross Layer Design, Rennes, France*, 30 November - 1 December 2011.

13. M Wódczak. Autonomic Cooperation in Ad-hoc Environments. *5th International Workshop on Localised Algorithms and Protocols for Wireless Sensor Networks (LOCALGOS) in conjunction with IEEE International Conference on Distributed Computing in Sensor Systems (DCOSS), Barcelona, Spain*, 27-29 June 2011.

14. M Wódczak. Convergence Aspects of Autonomic Cooperative Networking. *IEEE Fifth International Conference on Next Generation Mobile Applications, Services and Technologies, Cardiff, Wales, UK*, 14-16 September 2011.

15. M. Wódczak. Deployment Aspects of Autonomic Cooperative Communications in Emergency Networks. *3rd International Congress on Ultra Modern Telecommunications and Control Systems, IEEE ICUMT, Budapest, Hungary*, 5-7 October 2011.

16. M Wódczak. Resilience Aspects of Autonomic Cooperative Communications in Context of Cloud Networking. *IEEE First Symposium on Network Cloud Computing and Applications, Toulouse, France*, 21-23 November 2011.

17. M. Wódczak. Convergence Aspects of Autonomic Cooperative Networks. *International Journal of Information Technology and Web Engineering (IJITWE)*, 2012. Accepted for publication.

18. M. Wódczak. Cooperative Re-Routing. *full patent application no. 13/042701 filed to the United States Patent and Trademark Office*, 8 March 2011.

19. M. Wódczak, T. B. Meriem, R. Chaparadza, K. Quinn, B. Lee, L. Ciavaglia, K. Tsagkaris, S. Szott, A. Zafeiropoulos, B. Radier, J. Kielthy, A. Liakopoulos, A. Kousaridas, and M. Duault. Standardising a Reference Model and Autonomic Network Architectures for the Self-managing Future Internet. *IEEE Network*, 25(6):50–56, November/December 2011.

Chapter 8
Conclusion

In this book, the evolution path of spatio-temporal processing towards autonomic cooperative networking was presented. It was possible because over time, the rationale behind spatio-temporal processing was being gradually mapped onto networked systems. As a result, for example, initially the notion of the Multiple Input Multiple Output channel was transposed to Virtual MIMO and then further extensions were incorporated, such as relaying, as well as specific enhancements proposed by the author in the field of exploiting routing mechanisms and autonomic system design principles. To provide the most comprehensive and consolidated outline, this book first explained the rationale behind spatio-temporal processing, and then described the means of its mapping onto cooperative transmission. To this end, the concept of cooperative relaying over Virtual Antenna Arrays was employed, where each tier of Relay Nodes was emulating the operation of Distributed Space-Time Block Coding. For additional context, an analysis of adaptive approach to conventional relaying of the Manhattan type was presented. Following, an extension to cooperative relaying was proposed in the form of Routing information Enhanced Algorithm for Cooperative Transmission, which employed the mechanisms of the Optimised Link State Routing protocol for the purposes of organising the aforementioned Virtual Antenna Array aided and Space-Time Block Coded cooperative transmission. In particular, the Multi-Point Relay station selection heuristic was exploited to facilitate the integration of Virtual Antenna Arrays into the Optimised Link State Routing protocol. Although this integration was almost seamless, a group of modifications to the Optimised Link State Routing protocol was proposed to enable the new concept and to guarantee its backward compatibility. Since this solution could be perceived as more node-centric, it was further brought into a broader picture of the Generic Autonomic Network Architecture. For this reason the notion of Autonomic Cooperative Node was introduced, as well as both the Cooperative Transmission Decision Element and Cooperative Re-Routing Decision Element were incorporated into the autonomic cooperative system design. Following, the proposed extensions were applied to Relay Enhanced Cell to analyse the applicability of cooperative autonomic network deployments through the selection of cooperating Fixed Relay Nodes, chosen out of the mesh of Radio Access Point.

Such an approach showed advantages but the fixed nature of the scenario posed certain limitations. Eventually, this idea was applied to emergency communications networks where additional flexibility is available thanks to the existence of external Mobile Emergency Operations Centre, as well as Chief First Responders able to act as Mobile Relay Nodes.